BURT FRANKLIN: RESEARCH & SOURCE WORKS SERIES 773
Philosophy Monograph Series 66

PURE LOGIC

AND

OTHER MINOR WORKS

PURE LOGIC

AND

OTHER MINOR WORKS

BY

W. STANLEY JEVONS

EDITED BY ROBERT ADAMSON

AND

HARRIET A. JEVONS

WITH A PREFACE BY PROFESSOR ADAMSON

BURT FRANKLIN
NEW YORK

BC
135
J4
1971

Published by LENOX HILL Pub. & Dist. Co. (Burt Franklin)
235 East 44th St., New York, N.Y. 10017
Originally Published: 1890
Reprinted: 1971
Printed in the U.S.A.

S.B.N.: 8337-18428
Library of Congress Card Catalog No.: 71-160420
Burt Franklin: Research and Source Works Series 773
Philosophy Monograph Series 66

Reprinted from the original edition in the Wesleyan University Library.

PREFACE

IN the present volume are contained some of the earliest and some of the latest contributions of the author to the science of logic. The several works and papers that have been included fall into two distinctly marked groups, corresponding on the whole to the difference in time of their composition, and presenting at first sight no very clear indication of common aim or principle. The first group consists of independently published works, the *Pure Logic* and the *Substitution of Similars*, with certain papers in which the general principles of the author's logical theory, stated in their most mature form in his main logical works, the *Principles of Science* and the *Studies in Deductive Logic*, are carried out in certain special directions. In the second group appear such portions of a general examination of J. S. Mill's philosophy as had been completed at the date of the author's death.

The essays in the first group seem to claim a place in a permanent collection of the author's works, partly from the historic importance that must always attach to them, partly by reason of the relative fulness of discussion which they extend to certain fundamental and characteristic ideas in the general view of logic contained in them. They are reprinted as they originally appeared, with only the correction of one or two obvious misprints. In the author's copy of the *Pure Logic* an alteration was introduced in his own hand on § 146. For the expressions there given of De Morgan's two types of propositional form, viz.—

Everything is either A or B $A = b$
Some things are neither A nor B $a = b.$

There are substituted— $UA = b$
                        $Ua = bU.$

It has not been thought desirable to introduce into the text these alterations.

The arrangement of the four essays is chronological. *Pure Logic or the Logic of Quality Apart from Quantity, with Remarks on Boole's System, and on the Relation of Logic to Mathematics* was published in 1864. *The Substitution of Similars, the True Principle of Reasoning, Derived from a Modification of Aristotle's Dictum,* was published in 1869. The memoir *On the Mechanical Performance of Logical Inference* was received by the Royal Society 16th October 1869, and read 20th January 1870. The paper *On a General System of Numerically Definite Reasoning* was communicated in 1870 to the Literary and Philosophical Society of Manchester.[1]

There have not been included in this collection one or two contributions of less extent to logical science, the contents of which have been reproduced in more complete fashion in other works of the author. A communication to the Manchester Literary and Philosophical Society on 3d April 1866 deals with the *Logical Abacus,* which is elaborately explained in the later work, *Substitution of Similars.* A paper *On the Inverse, or Inductive, Logical Problem* (Memoirs of the Literary and Philosophical Society of

---

[1] As regards some points in this paper, see *Principles of Science*, second edition, p. 172.

Manchester, 1871-72), is incorporated with no substantive alteration in the *Principles of Science*, 1877, pp. 137-143. A short paper entitled *Who Discovered the Quantification of the Predicate?* appeared in the *Contemporary Review*, May 1873, pp. 821-824. It emphasises strongly the claims of Mr. George Bentham to recognition as the first English writer on Logic who had stated and applied the principle of Quantification. The paper is a comment on an article by Professor T. S. Baynes in the same journal, and as it formed part of a somewhat extended controversy, it has been judged needless to reprint it here.

The author himself attached so much weight to his critical examination of J. S. Mill's doctrines, and the labour bestowed on it played so large a part in the last ten or twelve years of his life, that the editors greatly regret their inability to add much to the four papers published during the author's lifetime in the *Contemporary Review*. From the large mass of MS. material for the further portions of the work, a general idea may be gathered as to the special lines of criticism that would have been followed; but with one or two exceptions no portion seemed to be in such a state that the editors without hesitation could print it as expressing the author's views. The four papers published by the author himself appeared in the *Contemporary Review*, Dec. 1877, Jan. 1878, April 1878, and Nov. 1879, under the title *John Stuart Mill's Philosophy Tested*. The fragment from the author's MSS. on *The Method of Difference* deals with the point referred to by him in the third of these articles. It is proposed here to indicate the relation in which the criticism of Mill stands to the author's logical theory, and to give a brief connected account of the lines it was to have followed.

The distinguishing feature of the doctrine first expounded

in the *Pure Logic* is the restriction of logical treatment to the qualitative aspect of thoughts. It was in this that the author found his difference from the system of Boole to consist; and from this follow the main features that characterise his logical method as a whole. There are certain consequences fairly involved in it, moreover, which though not formally stated by the author himself, are yet implied in his general theory of logic, and from which the precise ground of his emphatic dissent from Mill may be made tolerably clear. There is so much of agreement between the views of Mill and those expressed by the author in regard to the ultimate foundation of knowledge, both seeming to follow the lines of English empirical philosophy, that the strenuous dissent of the author from Mill's whole doctrine of logic at first presents a problem. But if the conception of a logic of quality be followed out, it will be found to involve an opposition between pure or formal logic and the theory of empirical or concrete reasoning, which for Mill constitutes logic, so sharp as to be irreconcileable. One or the other must yield. Briefly, it is the first that has pre-eminence in Jevons's view, the second is the all-important with Mill. Only within the limits of the first is there, according to Jevons, reasoning characterised by cogency and universality, but within its limits falls no determination of concrete existence; all its contents are, in reference to concrete existence, hypothetical. According to Mill there is no other field for reasoning than that of concrete existence; a pure logic is but a subordinate and relatively valueless abstraction from the actual processes of concrete reasoning.

To restrict reasoning, with its elements, the proposition and term, to quality is a position which seems to connect itself naturally with the familiar logical distinction between the comprehension and extension of notions. In his initial statements (*Pure Logic*, §§ 1, 3, 4) the author refers to this distinction, and though his expressions are not entirely in accordance with the current logical doctrine, it is easy to understand his position as resting upon it, and to treat his system as an effort to develop logical laws from the side of comprehension. But it may fairly be argued, both from the nature of the case and from many expressions of the author himself, that his system is wholly independent of that distinction, and indeed proceeds from a point of view which renders the distinction meaningless or inapplicable. The distinction as ordinarily expressed makes reference more or less explicitly to classification, and whether the basis be nominalist or conceptualist, leads the logician to contrast terms or notions which possess both comprehension and extension with those which are wanting in one or other aspect. It would seem impossible to work out consistently from this point without assigning to extension a certain measure of concrete existence, which, when closely investigated, has always been found to involve difficulty. Now it is abundantly evident that from all the implications of the view which finds expression in the distinction of comprehension from extension, the author's fundamental positions are entirely free. In his development of logic it is wholly superfluous to recognise the obscure and baffling differences that appear in the exposition of the various kinds of terms. Whatever be their origin, whether grammatical or metaphysical, these differences rest on concrete details of matter, and not on anything in the pure form of thought. The familiar distinctions on which the logician dwells, between Proper or Singular Terms and Common Terms, are of necessity rejected on the ground that quantitative difference is a secondary and derived aspect, not affecting the fundamental laws of thought, and dependent in each case on the special data implied. In a similar fashion, when the proposition is reached, it becomes evident that what it has to convey is altogether independent of the currently accepted distinctions of universal and particular, and that if these be retained at all, they must be interpreted in accordance with the fundamental laws of thinking that concerns itself with the identity or non-identity of the characterising qualities.

In adopting such a point of view, the logician is by no means prohibited from using forms of expression that seem to imply reference to concrete existences, to classes or single objects marked by the possession of qualities; and it is a further question, on which the author has hardly touched, whether it is possible to formulate the fundamental axioms of thinking in terms which do not imply such hypothetical objects. What is fundamental in his view is that the objects so referred to are hypothetical merely, are determined by the qualities thought as characterising them, and that they are absolutely to be distinguished from real concrete existences. On this account he emphasises so strenuously the position he shares with other logicians who have advanced to the same conclusion from widely different points of view, that in logic there is no question of existence,[1] and insists that logically non-existence is equivalent to contradictoriness. Such *existence* as may be assigned in logic to an object or class of objects is invariably possible existence. Only on material data then, and with reference to a defined universe of objects, could it be concluded that the universe possessed one quality or combination of qualities to the exclusion of its contradictory.

That it is very difficult to retain the position thus taken may at once be admitted, and it is evident that the author at various points of his exposition felt the difficulties. The

[1] *Studies in Deductive Logic*, pp. 141, 142.

particular proposition is an awkward adversary; there is some obscurity in the section of the *Pure Logic* (§ 146), where De Morgan's types of propositions are expressed in qualitative fashion; and a little hesitancy is observable in stating and defending the general criterion of consistency which is a legitimate and inevitable consequence of the view (*Pure Logic*, § 159, note; *Studies*, p. 181). But it does not seem that he ever wavered in regard to the fundamental position, that logically all judgments are non-existential.

According to this view, then, there is conceivable a pure logic, the complete statement of the ultimate conditions to which apprehension of qualities, as the same or different, is subject, and of the consequences that flow necessarily from such apprehension. Within the realm of pure logic we have absolute security and certainty. The laws to which thought necessarily conforms in dealing with identical or non-identical contents presented to it are at once the primary and the ultimate tests of truth. From these laws whatsoever absolute truth is attainable must be derived, and absolute certainty can only be assigned to deductions from them. Whenever we pass from the region of pure thought to regions of more concrete matter, we find that such measure of absolute certainty as may there be attainable depends on the completeness of conformity between the most general laws of such matter and the laws of pure thought. The elementary axioms of abstract arithmetic exhibit a perfect correspondence with the laws of pure thought, and are indeed derivatives from these laws. Number rests on logical discrimination or discrimination by pure thought, and its typical forms and laws are derivable from the conditions of pure thought when applied to the abstract schema of difference. The author has nowhere discussed with such fulness as the nature of the question demands, the connection between geometrical axioms and the laws of thought, and the grounds on which are to be rested the certainty and universality of geometrical reasoning.

There follows from the view thus described an immediate consequence which, when expressed as a logical doctrine, has become familiarly associated with the author's name. All reasoning, or, more exactly, all proof is deductive in character, and involves general propositions of absolute certainty. Proof is the more exact and appropriate term, for it remains within the limits of logical science or pure thought; whereas the term reasoning is vaguely extended to cover the natural, psychological process of arriving from data at a conclusion,—a process, therefore, of natural fact, about which the psychologist, not the logician, has to concern himself. The generality of the principle involved in any proof is not to be construed after the concrete fashion, as an assertion found to hold good about a number of concrete, and possibly not exhausted, particulars. It is either the quasi-concrete expression of an established identity of qualities, or the summary expression of exhaustive enumeration of a strictly limited collection of instances. It is either a qualitative or a numerical identity. In strictness, indeed, no concrete or synthetical proposition which is more than a collective expression for a determinate number of observed cases, can ever possess certainty. Whenever we pass beyond the pure laws of thought, with their numerical derivatives, or beyond the statement of actually observed fact, we are in the region of assumption and probability. The utmost that can be achieved is to determine with what degree of probability the general assumptions we make can be held. The rules for estimating such probability are numerical in character, and possess all the certainty of the laws of thought.

The range of assured knowledge is thus of narrow extent as compared with the indefinite expanse of concrete existence. But within that range it possesses absolute certainty. There and there only is proof possible. So soon as the consideration of concrete existence enters in, we are dependent on the hazardous data of intuitive experience, as to which we can never feel assured that they contain all that is required for complete insight. Within the region of concrete existence, reasoning, in the strict sense, is impossible. All apparent reasoning in which data and conclusions are assertions about concrete fact is hypothetical in character, and the hypotheses made, however various in concrete expression, are essentially the same in all cases,—viz. assumptions that the concrete propositions correspond perfectly to the conditions of pure abstract thought. It is impossible for us to know accurately that they do correspond; but, knowing the types of valid inference, we can gauge by the rules of probability the extent to which they approximate to the desired correspondence. Induction, then, as ordinarily described, is not a special mode of reasoning. Psychologically, we may no doubt pass in thought without further query from isolated particulars to a generality; but particulars can in no way substantiate what is not contained in them. There is only one method of reasoning—the deductive,—and it is used in concrete material as in abstract thought. But in the former case we have to note that our data involve assumptions that cannot be completely justified, and that are only more or less probable in a partially determinable degree.

Now it is the pointed opposition which the author makes between the narrow but secure region of pure, abstract thought and the wide sphere of more or less probable assumptions regarding concrete existence, an opposition involved in and springing from his conception of pure logic, that constitutes the real difference between his views and those of Mill. In the long run both agree in respect to the character of what is vaguely called 'knowledge' of concrete existence. Neither accorded absolute certainty to the propositions composing that 'knowledge.' Though Mill undoubtedly uses language in respect to so-called 'knowledge' of nature that implies the possibility of attaining certainty there, yet it may fairly be urged that such expressions are to be taken with the qualifications necessitated by the general doctrine, that knowledge of nature can never amount to absolute certainty. But, in Mill's view, this 'knowledge' of nature constituted all our knowledge, and within the realm of concrete existence lay the province and process of reasoning. Apparently abstract propositions, such as those of mathematics, were only abstracted from the concrete particulars, and had no other foundation for their certainty than the concrete experience from which they were drawn, no superior generality than followed from the relatively greater ease with which they could be disentangled from concrete details. The laws of pure thought, in which Jevons found the rules of absolutely certain knowledge and reasoning, were by Mill regarded as true but rather valueless prescripts defining the use of language, and having no function of significance when divorced from concrete fact. A pure logic, or logic of consistency in the employment of language, he did, indeed, admit to be possible, but in no way accorded to it special importance or viewed it as more than a result of abstraction from actual concrete reasoning. The possibility of an independent foundation for it he did not so much deny as ignore.[1]

The difference is as complete as could well be. To the

[1] The most explicit utterance of Mill on these points, which he rarely discusses, are in the *Examination of Hamilton* (third edition), pp. 457, 461.

one thinker, the theory of reasoning was a well-rounded, independent whole, narrow it might be in extent, but resting on principles of absolute certainty, not necessarily connected with any hypothesis as to the way in which knowledge of concrete fact is obtained or increased, and supplying the final standard by which evidence in concrete matters is to be tested. To the other, the theory of reasoning was part of the general doctrine of the ways in which an intelligence receiving its data as isolated empirical particulars gradually advanced to knowledge, to the establishment of general propositions, and to criticism of the grounds on which these rested. According to the first view, reasoning, in the strict sense of the term, was confined to the region of analytical thinking, and what accompanies reasoning,—viz. stringent cogency of proof, absolute certainty of conclusion, was possible only within that region. According to the other view, reasoning had little or no significance save in synthetical thinking, and its characteristics were such as could be obtained under the conditions of synthetical thinking.

It was inevitable, then, that from the author's point of view in logical theory, the apparent attempt in Mill's *System of Logic* to show how general knowledge in concrete material is gained and established, should present itself as a logical fallacy, doomed from the outset to failure, and only securing the outer aspect of coherence by skilful concealment of underlying inconsistency. For by knowledge the author understood what is perfectly conformable to the pure laws of thought and is warranted completely by them—attainable, therefore, only in analytical thinking. To admit, then as Mill seemed to do, the ultimate uncertainty of what is called knowledge of nature, and, while ignoring the pure laws of thought, to work out a theory of evidence on which knowledge might rest, was to occupy a wholly untenable position.

As regards the general point of view adopted in the proposed examination of Mill's philosophy, the motives with which it was undertaken, and the end aimed at, the author has given full explanation in the first of the series of articles contributed to the *Contemporary Review*. Among his MSS. are found portions of the sections intended to serve as the general introduction to the whole and as a summary conclusion. These contain in somewhat more detail the explanatory matter already given in the first article, and also some indication as to the arrangement of topics to be followed in the completed work. It would appear that after the general introduction the author purposed dealing in succession with Mill's *Essay on Religion*, with his views on Free Will and Necessity, with his peculiar reformation of the Utilitarian Ethics, with his doctrine of Inseparable Association, and then with the main object of attack, the Logical Theory. The criticism of the logic was the central, the fundamental portion of the work; the other discussions were given as illustrative confirmation of the estimate formed by the author, that Mill's mind was essentially illogical. It cannot however be determined, from the MSS., in what way these sections would have been arranged, nor is there any indication given as to the final order in which the various topics falling under logical theory would have been taken.

Taking first into consideration the sections not specifically logical, we find among the MSS. collected materials, with occasional written out portions of statement, relating to Mill's *Essay on Religion*, to his view on Free Will, and to his doctrine of Inseparable Association. As regards the *Essay on Religion*, it is clear from the fragments, as also from a reference to the topic in the first article in the *Contemporary Review*, that the author's intention was to dwell upon and enforce the apparent inconsistency between the two passages quoted from the essay in the said article.[1]

The section on *Free Will and Necessity* begins by a fairly written out statement of Sir William Hamilton's familiar position, that either conception, Liberty or Necessity, leads when developed to an ultimate inconceivability, a position which, on the whole, seems to have been accepted by the author. The remaining MSS. on the subject, in fragmentary remarks and references to passages in Mill's writings, enable us to see the drift of his argument, that Mill, while professing the Determinist view, makes admissions which are wholly irreconcilable with it and practically nullify it. The specific admissions singled out for closer scrutiny are two in number, closely connected, the one more psychological, the other more metaphysical in character. The first is the view expressed most definitely in the *System of Logic* (Book vi, chap. ii), that while the actions of the individual can only be regarded as the outcome of his character, desire to modify that character must be recognised as a factor in its formation. The second is contained in the distinction drawn in the posthumous Essays [2] between Nature as the entire system of things with all their attributes and Nature as that which is apart from human intervention. The author's intention seems to have been to press the arguments that, according to the principles on which Mill proceeds, desire to modify character must be treated like all other desires as a *derivative* and *determined* fact, not as original and determining, and that, even if for one purpose man be severed from nature

---

[1] *Three Essays on Religion*, p. 109. 'The essence of religion is the strong and earnest direction of the emotions and desires towards an ideal object, recognised as of the highest excellence, and as rightfully paramount over all selfish objects of desire.' *Ibid.* p. 103—' Religion, as distinguished from poetry, is the product of the craving to know whether these imaginative conceptions have realities answering to them in some other world than ours.'

[2] P. 46.

and regarded as intervening in it, on Mill's general principles, his activity of intervention can be contemplated in no other light than as a fact subject to natural law. If any other significance be assigned to the desire or intervening activity, the only result is utter inconsistency of theory. Mill's ethical theory is criticised, so far as the statement of Utilitarian doctrine is concerned, in the fourth of the articles published in the *Contemporary Review*. The MS. fragments entitled *Morals* indicate that the author purposed also discussing Mill's view on the relation between Intention and Motive, but do not suffice to yield a fairly precise statement of the points aimed at.

Under the heading *Metaphysics* a criticism is projected of the explanation Mill offers of Necessary Truths. The fragments show the author's intention to have been to examine the doctrine of Inseparable Association with special reference to the fundamental axioms of number, with a view to exposing the inconsistency between the character allowed to these axioms by Mill and the theory of their empirical origin through inseparable association.

A very large portion of the MSS. is concerned with Mill's logical theory. The several sections indicate, no doubt, the arrangement of topics to be made in the final treatment. We have been unable to find a definite statement as to this arrangement, and the order in which they are here referred to rests only on occasional expressions of the author. On the whole they fall into two main groups, those dealing with Mill's theory of the syllogism, those which criticise the doctrine of causation and the experimental methods. To the first group belong the sections headed *System of Logic, Petitio Principii, General Propositions, Particulars to Particulars*. The fragments forming these sections are evidently of various dates, and so far as any fairly written out portions are concerned, they frequently overlap one another. The following seems to have been the main line of criticism :—

Beginning with a pointed reference to the extreme inconsistency between the admitted novelty of Mill's view of Syllogism and the disclaimer of any novelty made in the Preface to the *System of Logic*, the author dwells on the general conception of the function of Logic as expressed by Mill. 'Logic is the Science of Proof or Evidence. In so far as belief professes to be founded on proof, the office of Logic is to supply a test for ascertaining whether or not the belief is well founded.' 'Logic neither observes, nor invents, nor discovers; but judges.'[1] Agreeing on the whole with this conception of the function of logic, but indicating a doubt as to whether even within the limits of the introduction to the *System of Logic*, Mill does not use expressions irreconcilable with it, the author proceeds to say that he will begin by showing that 'Mill upholds at one and the same time the four following doctrines :—(1) Logic is the science of Proof or Evidence. (2) The Syllogism is the only mode of reasoning by which we can assure ourselves of the correctness of an inference. (3) The Syllogism is nevertheless entirely optional, and imposed by the "arbitrary fiat" of logicians. (4) The Syllogism is at the same time necessarily a fallacy of the kind called Petitio Principii.'[2]

Turning to the fourth of these doctrines the author states fully Mill's familiar position, and urges in opposition to it that the conclusion is not contained in the major premiss but only in the major and minor combined. At the same time he recognises that such an answer by no means exhausts the question, and that the real foundation for Mill's view of

[1] *System of Logic* (seventh edition), vol. i, pp. 8, 9.
[2] From the author's MS.

the syllogism is to be sought in his treatment of the general proposition. From the passages selected for comment and from the fragments of criticism on them, it is evident that the author intended to insist that Mill regards the general proposition in two aspects, distinct and wholly irreconcilable, and that his inconsistency in at once rejecting the syllogism and allowing it to be the only way in which we can assure ourselves of the correctness of inference arises from the fact that one aspect of the general proposition is brought to bear in the formal discussion of syllogism, while the other predominates in the treatment of Induction and of the Deductive Method. The general proposition, Mill insists, looking at it from one point of view, is no more than the particulars it contains; it is based on and summarises particulars, and into particulars it is resolvable. Viewed in this light, the general proposition as major premiss, whether its truth be known or assumed, can not be the ground from which the truth of a conclusion contained in it is established. The form of reasoning in which it is so used, the Syllogism, is clearly guilty of petitio principii. But the general proposition has another aspect, not less common in Mill's treatment. The general proposition, as a result of inference, and in this aspect it has its greatest importance for us, is more than a summary of observed particulars. It embodies 'inferences and instructions for making innumerable inferences in unforeseen cases.' Wherever an inference to a particular case is well founded, is valid, the conclusion is already a general proposition. Data sufficient to establish one instance are sufficient to establish a class of instances.[1]

If, then, reasoning or proof cannot legitimately advance

[1] The author evidently projected a special criticism on Mill's view of a *class*, the notion of which, as defined by Mill, seemed to him to involve the same ambiguity and inconsistency as the treatment of general propositions. The two questions are, indeed, one substantially.

from a general proposition, but if a general proposition may be established by reasoning, it becomes necessary to ask what, according to Mill, is the common form of reasoning, and how is the validity of a general conclusion established? Consideration of the first question is, in brief, criticism of Mill's well-known doctrine, that all reasoning is fundamentally inference from particulars to particulars; consideration of the second leads to a review of many salient points in Mill's theory of Induction.

The discussion of the view that inference is from particulars to particulars is incomplete. From what remains, it seems that the author would have pressed the argument that though, in point of fact, we may and do pass mentally from particulars to particulars, the process is itself a 'complicated and precarious combination of induction and deduction,' indefinitely far removed, therefore, from satisfactory proof or evidence, and the rules of which, if possible at all, must involve reference to considerations that go beyond particulars. In other words, admitting that reasoning from particulars to particulars may be an actual, psychical process, the author maintained that the identification of the conditions of a natural process with the conditions of proof was not only an error, but wholly inconsistent with Mill's general conception of logic as the science not of the ways in which we do reason but of the rules of valid reasoning.

Some portion of the criticism of the theory of Induction is contained in the third of the articles from the *Contemporary Review*, and in the fragment, printed in this volume, on the *Method of Difference*. What remains was arranged by the author under the heads *Baconian Method* and *Causation*. In reference to the first point, the author draws attention to the character of Mill's objection to the use of Induction as described by Bacon, and to his insistence on the necessity for general reasoning. In this, and generally in all that Mill has described under the head of the Deductive Method, the author finds what at once corrects and destroys the erroneous conceptions of reasoning as inference from particulars to particulars, and of Induction as a process whereby we arrive at general results from particular premisses.[1]

The section on *causation* was evidently intended to take up in succession the definition of cause and of the causal relation offered by Mill, the assumption of the universality of causation, and the grounds on which its universality was rested. In examining the definition of cause, the author draws attention to the ambiguity of Mill's language, which allows a two-fold interpretation to be easily put upon the all-important terms, invariable antecedent and consequent. An invariable sequence asserted might mean, he points out, either that

x invariably follows X,

or that while x invariably follows X,
x is invariably preceded by X.

The language employed, he insists, leaves the ambiguity unresolved, and indeed tends to favour the second mode of interpretation. It is only when, at a later stage, *plurality of causes* is formally introduced that we learn how abstract our interpretation of invariable antecedent must be. The author's MSS. contain no further criticism of this point, but his view on the subject can readily be gathered from his published works.[2]

On the second point, the universality of causation, the author contrasts at length passages in which the certainty of general knowledge is assumed, and is stated to depend on

[1] See on this point *Principles of Science* (second edition), pp. 265, 508.
[2] See *Principles of Science* (second edition), pp. 222, 226, 737 *sqq.*

the law of causation, with the passage [1] in which Mill asserts that 'the uniformity in the succession of events, otherwise called the law of causation, must be received not as a law of the universe, but of that portion of it only which is within the range of our means of sure observation, with a reasonable degree of extension to adjacent cases.'

As regards the third point, the treatment of the grounds for the assumption of a universal law of causation coincides throughout with the criticism of *Inductio per enumerationem simplicem* contained in the third of the articles on J. S. Mill's Philosophy.

To this brief and necessarily inadequate summary of the author's projected work, it must be added that only contact with the MSS. can convey any fair idea of the painstaking and conscientious manner in which the scrutiny of Mill's writings had been carried out. Whatever opinion may be formed of the value of the general objections taken to Mill's logical and philosophical doctrines, or of the appropriateness of the limits imposed by the author on his criticism, it must be acknowledged that he took every precaution against oversight or hasty judgment, and that his every utterance was supported by the fullest evidence attainable. The investigation of the fundamental principles of reasoning is a problem of such sublety and complexity that exhaustive criticism of one distinguished logician by another must always be hailed with satisfaction. It is not the least part of the severe loss which science and philosophy incurred by the author's untimely death, that he was prevented from utilising, as he only could do, the materials he had collected.

[1] *System of Logic*, Book iii, chap. xxi, sec. 4, concluding paragraph.

CONTENTS

PART I

WRITINGS ON THE THEORY OF LOGIC

	PAGE
I. PURE LOGIC OR THE LOGIC OF QUALITY APART FROM QUANTITY. (1864)	1
II. THE SUBSTITUTION OF SIMILARS. (1869)	79
III. ON THE MECHANICAL PERFORMANCE OF LOGICAL INFERENCE. (From the *Philosophical Transactions*, 1870)	137
IV. ON A GENERAL SYSTEM OF NUMERICALLY DEFINITE REASONING. (From the *Memoirs of the Manchester Literary and Philosophical Society*, 1870)	173

PART II

JOHN STUART MILL'S PHILOSOPHY TESTED

(*Portions of an Examination of J. S. Mill's Philosophy*)

I. ON GEOMETRICAL REASONING	199
II. ON RESEMBLANCE	222
III. THE EXPERIMENTAL METHODS	250
IV. UTILITARIANISM	268
V. ON THE METHOD OF DIFFERENCE	295

(I.—IV. from the *Contemporary Review*; V. from the Author's MSS.)

PART I

WRITINGS ON THE THEORY OF LOGIC

PURE LOGIC

OR THE

LOGIC OF QUALITY APART FROM QUANTITY

WITH REMARKS ON BOOLE'S SYSTEM

AND ON THE RELATION OF LOGIC AND MATHEMATICS

INTRODUCTION

IT is the purpose of this work to show that Logic assumes a new degree of simplicity, precision, generality, and power, when comparison in quality is treated apart from any reference to quantity. *Extent and intent of meaning.*

1. It is familiarly known to logicians that a term must be considered with respect both to the individual things it *denotes*, and the qualities, circumstances, or attributes it *connotes*, or implies as belonging to those things. The number of individuals denoted forms the breadth or *extent* of the meaning of the term; the qualities or attributes connoted form the depth, comprehension, or *intent*, of the meaning of the term. The extent and intent of meaning, however, are closely related, and in a reciprocal manner. The more numerous the qualities connoted by a term, the fewer in general the individuals which it can denote; the one dimension, so to speak, of the meaning being given, the other follows, and cannot be given or taken at will.

4 PURE LOGIC

Expression usually combined. 2. Logicians have generally thought that a proposition must express the relations of extent and intent of the terms at one and the same time, and as regarded in the same light. The systems of logic deduced from such a view, when compared with the system which may otherwise be had, seem to lack simplicity and generality.

Separation necessary. 3. It is here held that *a proposition expresses the result of a comparison and judgment of the sameness or difference of meaning of terms, either in intent or extent of meaning.* The judgment in the one dimension of meaning, however, is not independent of the judgment in the other dimension. It is only then judgment and reasoning in one dimension which is properly expressed in a simple system. Judgment and reasoning in the other dimension will be and must be implied. It may be expressed in a numerical or quantitative system corresponding to the qualitative system, but its expression in the same system destroys simplicity.

I do not wish to express any opinion here as to the nature of a system of logic in extent, nor as to its precise connection with the pure system of logic of quality.

Primary system. 4. Reasoning in quality and quantity, in intent or extent of meaning, being considered apart, it seems obvious that the comparison of things in quality, with respect to all their points of sameness and difference, gives the primary and most general system of reasoning. It even seems likely that such a system must comprehend all possible and conceivable kinds of reasoning, since it treats of any and every way in which things may be same or different. All reasoning is probably founded on the laws of sameness and difference which form the basis of the following system.

Present task. 5. My present task, however, is to show that *all and more than all the ordinary processes of logic may be combined in a system founded on comparison of quality only, without reference to logical quantity.*

Relation to Boole's Logic. 6. Before proceeding I have to acknowledge that in a considerable degree this system is founded on that of Professor Boole, as stated in his admirable and highly original Mathematical Analysis of Logic.[1] The forms of my system may, in fact, be reached by divesting his system of a mathematical dress, which, to say the least, is not essential to it. The system being restored to its proper simplicity, it may be inferred, not that Logic is a part of Mathematics, as is almost implied in Professor Boole's writings, but that the Mathematics are rather derivatives of Logic. All the interesting analogies or samenesses of logical and mathematical reasoning which may be pointed out, are surely reversed by making Logic the dependent of Mathematics.

[1] *Investigation of the Laws of Thought.* By George Boole, LL.D. London, 1854. Frequent reference will be made to this work in the following pages.

CHAPTER I

OF TERMS

Of things and their names.
7. *Pure logic arises from a comparison of things as to their sameness or difference in any quality or circumstance whatever.*

In discourse we refer to things by the aid of marks, names, or terms, which are also, as it were, the handles by which the mind grasps and retains its thoughts about things. Thus correct thought about things becomes in discourse the correct use of names. Logic, while treating only of names, ascertaining the relations of sameness and difference of their meanings, treats indirectly, as alone it can, of the samenesses and differences of things.

Meaning of name.
8. A term taken in intent has for its meaning the whole infinite series of qualities and circumstances which a thing possesses. Of these qualities or circumstances some may be known and form the description or definition of the meaning; the infinite remainder are unknown.

Among the circumstances, indeed, of a thing, is the fact of its being denoted by a given name, but we may speak of a thing, of which only the name is known, as having *a name of unknown meaning*.

The meaning of every name, then, is either unknown or more or less known. But we may speak of a term that is more or less known as being simply *known*.

Qualities infinite in number.
9. Among the qualities and circumstances of a thing is to be counted everything that may be said of it, affirmatively

or negatively. Any possible quality or circumstance that can be thought of either does or does not apply to any given thing, and therefore forms, either affirmatively or negatively, a quality or circumstance of the thing. Concerning anything, then, there may be an infinite number of statements made, or qualities predicated.

10. When we assign a name to a thing, with knowledge of, and regard to, certain of its qualities or circumstances, that name is equally the name of anything else of exactly the same known qualities and circumstances. For there is nothing in the name to determine it to the one thing rather than the other. Any name, then, must be the name *in extent* of anything and of all things agreeing in the qualities or circumstances which form its known meaning *in intent*, and in this system. *[Relation of meanings.]*

11. Though it is well to point out that all our names or terms bear a universal quantity when regarded in extent, it must be understood, and constantly borne in mind, that further reference to the meaning of a term in extent or quantity of individuals, is excluded in these pages. *[Present meaning.]*

The primary and only present meaning of a name or term is a certain set of qualities, attributes, properties, or circumstances, of a thing unknown or partly known.

12. *Term* will be used to mean *name*, or any combination of names and words describing the qualities and circumstances of a thing. *[Term defined.]*

13. The terms of this system may be made to express any combination of samenesses and differences in quality, kind, attribute, circumstance, number, magnitude, degree, quantity, opposition, or distance in time or space. A term may thus represent the qualities of a thing or person in all the complexity of real existence, so well and fully defined that we cannot suppose there are, or are likely to be, two things the same in so many circumstances. Such a term would correspond to the *singular, proper, non-attributive,* or *non-connotative* names of the old logic. Such names are accordingly by no means excluded from this system; and *[Generality of our terms.]* *[Proper names.]*

8   PURE LOGIC

it is here held that the old distinction of *connotative* and *non-connotative* names is wholly erroneous and unfounded. If there is any distinction to be drawn, it is that singular, proper, or so-called non-connotative terms, are more full of connotation or meaning in intent or quality than others, instead of being devoid of such meaning.

Condition of sameness of meaning.

14. As logic only considers the relations of meaning of terms, as expressed within a piece of reasoning, the special meaning of any term is of no account, provided that *the same term have the same meaning throughout any one piece of reasoning.*

Thus, instead of the nouns and adjectives, to each of which a special meaning is assigned in common discourse, we shall use certain letters, A, B, C, D, ... U, V . . . each standing for a special term, or a definite meaning, and for *any term or meaning*, always under the above condition.

Terms known or unknown.

15. Our terms, A, B, C, ... like the terms of common discourse, may be either known or unknown in meaning. *It is the work of logic to show what relations of sameness and difference between unknown and known terms may make the unknown terms known.*

Were it not to explain *ignotum per ignotius*, we might say that *logic is the algebra of kind or quality, the calculus of known and unknown qualities*, as algebra (more strictly speaking *universal arithmetic*, which does not recognise essentially negative quantities) is the calculus of known and unknown quantities.

Symbols of plain meaning.

16. Let it be borne in mind that the letters A, B, C, etc., as well as the marks $+$, 0, and $=$, afterwards to be introduced, are in no way mysterious symbols. The term A, for instance, is merely a convenient abbreviation for any ordinary term of language, or set of terms, such as *Red*, or *the Lords Commissioners for executing the office of the Lord High Admiral of England.*

Again, $+$ is merely a mark substituted for the sake of clearness, for the conjunctions *and, either, or*, etc., of common language. The mark = is merely the copula *is*, or *is same as*, or some equivalent. The meaning of 0, whatever it exactly be, may also be expressed in words. There is consequently nothing more symbolic or mysterious in this system than in common language.

CHAPTER II

OF PROPOSITIONS

Proposition defined.
17. *A proposition is a statement of the sameness or difference of meaning of two terms*, that is, of the sameness or difference of the qualities and circumstances connoted by each term.

Affirmative, negative.
18. According as a proposition states *sameness* or *difference*, it is called *affirmative* or *negative*.

Its purpose.
19. *It is the purpose or use of a proposition to make known the meaning of a term that is otherwise unknown.*

Truth and falsity.
20. A proposition is said to be true when the meanings of its terms are same or different, as stated; otherwise it is called false or untrue. As logic deals with things only through terms, it cannot ascertain whether a proposition is true or false, but only whether two or more propositions are or are not true together, under the condition of meaning of terms (§ 14).

Notation of affirmative proposition.
21. We denote by the copula *is*, or by the mark =, the sameness of meaning of the terms on the two sides of a proposition.

For the present we shall speak only of affirmative propositions, which are of superior importance; and when not otherwise specified, *proposition* may be taken to mean *affirmative proposition*.

Conversion of propositions.
22. *A proposition is simply convertible.* The propositions $A = B$ and $B = A$ are the same statement; either of the terms A and B is the same in meaning as the other, undistinguishable except in name.

This simple conversion comprehends both the *simple conversion*, and *conversio per accidens* of the school logic.

23. *One proposition and one known term may make known one unknown term.* *One term known from one proposition.*

From A = B, so far as we know B, that is, know its meaning, we can learn A; so far as we know A, we can learn B.

We thus know samely of both sides of a proposition whatever we know of either. The same might be said of uncertain or obscure knowledge.

24. A proposition between any two terms of which the meanings are otherwise known as same or different, is useless. For it cannot serve the purpose of a proposition (§ 19). Such is any proposition between a term and itself, as A = A, B = B (§ 14). These useless propositions are called *Identical*. They state the condition of all reasoning, but we know it without the statement. *Useless and identical propositions excluded.*

A proposition repeated, or a converted proposition (§ 22), is also useless, except for the mere convenience of memory, or ready apprehension.

CHAPTER III

OF DIRECT INFERENCE

Law of sameness.
25. It is in the nature of thought and things, that *things which are same as the same thing are the same as each other.*

More briefly—SAME AS SAME ARE SAME.

Hence the first law of logic—that *terms which are same in meaning as the same term, are the same in meaning as each other.*

This law, it is obvious, is analogous to Euclid's first axiom, or common notion, that *things which are equal to the same thing, are equal to each other.* Things are called equal which are *same in magnitude,* but what is true of such sameness, is also true of sameness in any way in which things may be same or different. Euclid's geometrical law is but one case of the general law.

Meaning of laws of logic.
26. Logic proceeds by laws, and is bound by them. For logic must treat names as thought treats things. And the laws of logic state certain *samenesses* or uniformities in our ways of thinking, and are of self-evident truth.

Direct inference shown and defined.
27. *When two affirmative propositions are same in one member of each, the other members may be stated to be same.*

From $A=B$, $B=C$, which are the same in the member B, we may form the new proposition, $A=C$. For A and C being each stated to be the same as B, may by the law of sameness be stated to be the same as each other.

A proposition got by the Law of Sameness is said to be got by *direct inference,* and is called a *direct inferent,* or, in common language, *a direct inference.*

DIRECT INFERENCE

28. Propositions from which an inference is drawn are called *premises*, and are given or taken as the basis of reasoning. Logic is not concerned with the truth or falsity of premises or inference, except as regards the truth or falsity of one with the other (§§ 20, 37). *Premise defined.*

29. An *expression* for a term consists of any other term which by premises we know to be the same in meaning with that term. *Expression defined.*

30. In inferring a new proposition from two premises we are said to *eliminate* or remove the member which is the same in the two premises. *Elimination defined.*

From two premises we may eliminate only one term, and infer one new proposition. By saying that *we may*, it is not meant that *we always can*.

31. Propositions are said to be *related* to each other which have a same or common member, or which are so related to other propositions so related; and so on. *Related propositions defined.*

In other words, any two propositions are related which form part of a series or chain of propositions, in which each proposition is related to the adjoining ones or one.

32. Terms are said to be *related* which occur in one same, or in any related propositions. *Related terms defined.*

33. *From two related premises and one known term we may learn two unknown terms, and not more.* *Use of syllogism*

From $A = B$ and $B = C$, we learn any two of A, B, C, when the third is known.

34. *From any series of related premises, and one known term we may learn as many unknown terms as there are premises.* Thus from $A = B = C = D = E = F$, we may learn any five terms when the sixth is known. For each useful proposition may render one unknown term known (§ 19). Between each two adjoining premises one term may be eliminated, becoming known in one premise, and rendering another term known in the other. There must at last remain a single proposition containing two terms, each of which occurs only in one premise. *Series of premises.*

35. *The number of related premises must be one less than*

Number of terms and related premises. the number of different terms. If it be still less, the propositions cannot be all related; if it be greater, some of the premises must be useless, because they must lie between terms otherwise known to be same by inference.

It will be remarked that systems of mathematical propositions or equations with known and unknown quantities are perfectly analogous in their properties to logical propositions.

Irrelevant terms and premises. **36.** When a related premise contains a term or member not relevant to the purpose of the reasoning, this term is eliminated by neglecting the premise; and for every such premise neglected a term is eliminated. In regard both to related and unrelated premises and their terms, the *neglect* of all irrelevant terms and premises may be considered a process of elimination which accompanies all thought.

Science of Science. **37.** *Inference is judgment of judgments, and ascertains the sameness of samenesses.*

When in comparing A with B, and the same B with C, we judge that $A = B$ and $B = C$, we obtain *science*, or reasoned knowledge of things, as distinguished from the mere knowledge of sense or feeling. But when we judge the judgments $A = B = C$ to be the same, as regards A and C, with the judgment $A = C$, we obtain *Science of Science*.

Here is the true province of logic, long called *Scientia Scientiarum*. Hence it is that logic is concerned not with the truth of propositions *per se* (§ 20), but only with the truth of one as depending on others.

SCIENCE OF SCIENCE	$\{A = B = C\} = \{A = C\}$		REASONING
SCIENCE	$A = B$	$B = C$	JUDGMENT
THINGS	A	B	C APPREHENSION

Form and matter. **38.** Here we find the clear meaning of the distinction of *form and matter of thought.*

$$\left. \begin{array}{rcl} \text{Sameness of Samenesses} & = & \text{Form} \\ \text{Sameness of things} & = & \left\{ \begin{array}{c} \text{Matter} \\ \text{Form} \end{array} \right. \\ \text{Things} & = & \text{Matter} \end{array} \right\} \text{Of thought.}$$

CHAPTER IV

OF COMBINATION OF TERMS

39. In discourse, when several names are placed together side by side, the meaning of the joint term is sometimes the sum of the meanings of the separate terms.[1]   *Addition of meanings.*

So in our system, *when two or more terms are placed together, the joint term must have as its meaning the sum of the meanings of the separate terms.* These must be thought of together and in one.

40. Any terms placed together will be said to form with respect to any of those separate terms a *combination* or *combined term*. With respect to all other terms they may be called simply *a term*. For it must be remembered that any single term, A, B, C, etc., is not more single in meaning than a combination.   *Combination defined.*

---

[1] I shall here consider only the cases of combination in which the combined term means the *added meanings* of the separate terms. The same forms of reasoning apply, as I believe, *mutatis mutandis*, to any cases of combination under some such wider law as this—

Same parts samely related make same wholes.

Only by some such extension can logic be made to embrace the major part of all ordinary reasoning, which has never yet been embraced by it, save so far as this may have been done in some of Professor De Morgan's latest writings. But to show how such an extension may be grafted on to my system must be reserved for a future opportunity. In most relations it is obvious that the order of terms in relation is no longer indifferent (§ 41).

Concerning some inferences by combination, see Thomson's *Outlines*, §§ 87, 88.

Order of combination indifferent.

41. *The meaning of a combination of terms is the same in whatever order the terms be combined.*

Thus $AB = BA$; $ABCD = BACD = DCAB$, and so on.

For the order of the terms can at most affect only the order in which we think of them, and in things themselves there is no such order of qualities and circumstances (Boole, p. 30).

Law of simplicity.

42. *A combination of a term with itself is the same in meaning with the term alone.*

Thus $AA = A$, $AAA = A$, and so on.

Also, a combination of terms is not altered by combination with the whole or any part of itself. Thus $ABCD = ABCD \cdot BCD = A \cdot BB \cdot CC \cdot DD = ABCD$, since $BB = B$, $CC = C$, $DD = D$.

The coalescence of same terms in combination must be constantly before the reader's mind.

This important and self-evident law of logic was first brought into proper notice by Professor Boole (p. 32), who remarks: 'To say "good, good," in relation to any subject, though a cumbrous and useless pleonasm, is the same as to say "good."'

Professor Boole gave to this law the name *Law of Duality*. But as this name, on the one hand, is not peculiarly adapted to express the general fact, $AAAAA\ldots = A$, and is peculiarly adapted to express the fact $A = AB + Ab$ (§ 99), I have ventured to transfer the name, and substitute a new one.

Degree of quality.

43. In the terms as used under the above law there is no reference to *degree of quality*. When required, each degree of quality may be treated in a separate term, containing as part of its meaning every less degree of the quality. Two or more degrees of a same quality in logical combination therefore produce the greatest of those degrees.

Law of same parts and wholes.

44. It is in the nature of thought and things that *when same qualities are joined to same qualities the wholes are same*.

Hence the law of logic—

COMBINATION

Same terms combined with same terms give same combined terms.

Thus, since $A = A$ and $B = B$, therefore $AB = BA = AB$.

This self-evident law is a more general case of Euclid's second axiom. It may, perhaps, be most briefly stated as follows:—*Same parts make same wholes.*

45. *Same terms being combined with both members of a premise, the combinations may be stated as same in a new proposition which will be true with the premise.* *Inference by combination.*

For what is true of terms obviously the same, as A, A, or B, B, must also be true of terms known to be the same in meaning by a premise. Thus, from $A = B$ we may infer $AC = BC$ by combining C with each of A and B.

As the number of possible terms which may be combined with the terms of a premise is infinite, there may be drawn from any premise an infinite number of inferences by combination.

46. Inferences which may be drawn by combining the members of two or more premises need not be considered here. *Combination of propositions.*

47. A proposition inferred by combination (§ 45) will be true with its premise, whatever be the term or terms used for combination. When terms of specific meaning, indeed, are selected at random, it will usually happen that the combinations of the inference are unheard-of, absurd, and useless. This does not affect the truth of the inferred proposition, which only asserts that the meaning of the one combination, whatever it be, is the same as the meaning of the other. *General truth of such inferences.*

48. In our daily use of specific terms, we constantly use each under the restriction of a number of premises so well known to all persons that it is needless to express them. Terms joined not in accordance with these tacit relations make nonsense. For instance, the impassable difference of matter and mind renders it nonsense to join the name of any material with that of any mental attribute, except in a merely metaphorical sense. In order, then, that our inferences should always be intelligible and useful, we *Tacit relations excluded.*

should require the expression of all tacit premises connected with terms of specific meaning. It is only the several branches of science, however, that can undertake the necessary investigations in detail. Our inference remains true, however complicated be the relations of sameness and difference of the terms introduced. But it is inference from premises which *are stated*, not from those which *might be* or *ought to be stated*.

Formation of common term.

49. *When premises contain terms only partially the same, the combination of each with the part that is different in the other will produce a term completely the same in each.* Such premises may be considered as *related* (§ 31).

Thus, in $A = C$ and $B = CD$, the terms C and CD are only partially the same. But the combination of D with $A = C$ gives $AD = CD$, having one member completely the same as one member of $B = CD$. Hence we may infer $AD = CD = B$ (§ 26), and eliminate the term C, which was common in the premises: thus, $AD = B$.

Again, to eliminate B from the premises $A = BC$ and $E = BD$, combine D with each side of the first, and C with each side of the second. Hence, $AD = BCD = CE$, or $AD = CE$, in which B does not appear.

Useless cases.

50. From premises which have no term in common, this process will only give us the inferences which might be had (§ 46) by the direct combination of the respective terms of the premises. Thus $A = B$ and $C = D$ give $AD = BD$, and $BC = BD$, whence $AD = BC$. And we might similarly get $AC = BD$.

Substitution defined.

51. The following process may be called *substitution*, and will be seen to give the same inference as the two processes of forming a common term (§§ 49, 27), and then eliminating it.

For any term, or part-term, in one premise, may be substituted its expression (§ 29) *in other terms.*

In short, the two members of any premise may be used indifferently, one in place of the other, wherever either occurs.

Thus, if $A = BCD$ and $BC = E$, we may in the former

premise substitute for BC its expression E, getting A = DE.
The full process of inference consists in combining D with
both sides of BC = E, and eliminating the complete common
term BCD thus obtained, so that A = BCD = DE.

52. *We may substitute for any part of one member of a* Intrinsic
proposition the whole of the other.
elimination.

Thus, in A = BCD, we may substitute for any one of
B, C, D, BC, BD and CD, parts of BCD the one member,
the whole, A, of the other member, inferring the new
propositions—

$$A = ACD \qquad A = ABD \qquad A = ABC$$
$$A = AD \qquad A = AB \qquad A = AC.$$

The validity of this process depends on the Laws of
Simplicity (§ 42), and of Part and Whole (§ 44), as is seen
by combining each member of the premise with itself.
Thus, from A = BCD we have A.A = BCD.BCD = BCD.D = AD
by coalescence of same terms, and substitution for BCD
of its expression A.

The new proposition thus inferred will have one of its
sides *pleonastic*, that is, with some part of its meaning
repeated. But it is obvious that we cannot, as a general
rule, substitute for part of one side less than the whole of
the other, because we cannot from the premise alone know
that the meaning of the part-term removed is quite supplied
in the part of the other member put for it.

The above process may be called *intrinsic elimination*, to
distinguish it from the former process of elimination between
two premises, which may be called *extrinsic elimination*, and
is seen to be that case of intrinsic elimination in which we
substitute for the whole of one side the whole of the other.
In a single premise, intrinsic elimination of a whole member
would give only an identical and useless result.

Intrinsic elimination gives no new knowledge, but is of
constant use in striking out or *abstracting* terms concerning
which we do not desire knowledge, and which are therefore
worse than useless in our results.

Professor Boole's system of elimination (p. 99), is, I believe, equivalent to the above, though the correspondence may not at first sight be apparent.

Failure of elimination.
53. A term cannot be intrinsically eliminated which occurs in both members of a proposition. The presence of such part-term may be called a *condition* of the sameness of the remainder of the terms.

Samely related terms.
54. Terms are said to be samely related in a premise when their interchange does not alter the premise.

Thus, B and C are samely related in A = BC, because the premise is the same, A = CB (§ 41), after their interchange. But A and B are not samely related, because their interchange alters the premise into B = AC.

In A = BCDE . . . any two of B, C, D . . . . are samely related and may be interchanged.

Inference concerning two such.
55. Of samely related terms, an expression for the one is the same as the expression for another after the two terms in question have been interchanged.

Concerning several.
56. When several terms are samely related, we obtain the expressions concerning the rest from the expression for any one by successively changing each term into the next when the terms are kept in some fixed order.

It is evident that we may always interchange the terms in any part of a problem, provided we do so throughout the problem (§ 14). And in those cases in which the premises remain unchanged thereby, we evidently get several inferences from the same premises. This method of interchanges is familiar to mathematicians.

Mathematical analogy.
57. It will be obvious that a mathematical term or quantity of several factors is strictly analogous in its laws to a logical combined term, *excluding the Law of Simplicity.*

CHAPTER V

OF SEPARATION OF TERMS

58. It is in the nature of thought and things that *when from same sets of qualities same qualities are taken, the remaining sets are the same;* or, more briefly—*Same parts from same wholes leave same parts.*

Law of same wholes and parts.

Hence the logical law :—*When from same combinations of terms same terms are taken, the remaining terms are the same.*

This is the converse of the Law of Same Parts and Wholes (§ 44), and is equally self-evident with it. But it is not equally useful with it; and in Pure Logic, in fact, is of no use at all. The removal of *terms* with their known meanings is not equally possible with their combination, and in useful logical premises, is not possible at all. For, in a useful premise (§ 19), a part at least of one member must be unknown, and this part may or may not contain the part we desire to remove. Even supposing then that a term occurs on either side of a premise, we cannot remove it from the known side, because we cannot know whether or not we can remove it from the unknown or partially known side.

Thus in $AB = AC$, suppose A and C known, and B unknown. We cannot infer $B = C$, because B may contain part or the whole of the known meaning of A, in addition to the known meaning of C, by the Law of Simplicity (§ 42), and in leaving B, we do not remove A from one member of the premise.

Complete analogy of logic and mathematics.

59. The logic of known and unknown terms, it has been said (§ 15), is analogous to the calculus of known and unknown numbers.

So, a logic in which all terms were known would have an analogue in common Arithmetic, a calculus in which all the numbers employed are known. Combination of terms has an analogue in multiplication of numbers, and separation of terms in division of numbers. As in logic combination is unrestricted, so in calculus is multiplication. As in logic of known terms only, separation of terms is unrestricted, so in a calculus of known numbers only, division is unrestricted. But, *as in logic of known and unknown terms separation is restricted, so in calculus of known and unknown numbers division is restricted.*

Restriction of division.

60. It is well known that, in like manner, we cannot divide both sides of an equation by an unknown factor, and assert the resulting equation to be necessarily true, because the unknown factor may be $=0$. Thus, from $xy=xz$, we cannot remove x, and assert $y=z$, because if x happen to be $=0$, the equation $xy=xz$ is true, whatever finite numbers be the meanings of y and z.

The correspondence is thus shown :—

Logical Propositions.	Mathematical Equations.
Terms known admit	*Numbers known admit*
Combination	Multiplication
Separation	Division
(unless either dividend contain divisor)	(unless divisor $=0$)
Terms unknown admit	*Numbers unknown admit*
Combination	Multiplication
but do not admit	*but do not admit*
Separation	Division.

The above analogies did not escape the notice of Professor Boole (pp. 36-37), and I am therefore at a loss to understand on what ground he asserts that there is a breach in the correspondence of the laws of logic and mathematics.

Parts not known from the whole.

61. *From the meaning of a whole term we cannot learn the meaning of a part.*

In $A=BC$, if we know A we learn BC as a whole; but

we do not thence learn the parts B, C, separately. For of the qualities in A any part may be in B, and any part in C, including any part of those in B, by the Law of Simplicity (§ 42). It is only necessary that every quality in A shall be either in B or in C. Even if we know one of B and C, we only learn of the other that it must contain any quality of A not in the first.

We here meet the imperfection of an inverse process.

62. With reference to the relation between the number of premises, and the numbers of known and unknown terms (§§ 33-35), we must treat as separate terms any which occur separate in premises, although they may also occur in combination. Otherwise, we always treat any whole combination as a single term.

Number of terms and premises.

CHAPTER VI

OF PLURAL TERMS

Terms of many meanings.
63. *A plural term has one of several meanings, but it is not known which.*

Thus *B or C* is a plural term, or term of many meanings, for its meaning is either that of B or that of C, but it is not known which.

A term not in form plural, may be distinguished as *single;* such is A.

Alternative defined.
64. The separate terms expressing the several possible meanings of a plural term are called *alternatives*, and are to be joined together by the sign + placed between each two adjoining terms.

All that has been said of single terms applies to plural terms, *mutatis mutandis*.

Order of alternatives.
65. *The meaning of a plural term is the same whatever be the order of the alternatives.*

Either B or C is the same in meaning as *either C or B*, that is, $B+C=C+B$. For the order in which we think of the possible qualities of a thing cannot alter those qualities, and the order must not convey any intimation that one meaning is more probable than another.

Combination with plural term.
66. *A term is combined with a plural term by combining it with each of its alternatives.*

For what is A and either B or C, if it is B, is AB; if it is C, is AC, and it is therefore either AB or AC.

Use of brackets.
67. Let a plural term enclosed in brackets (. . . . . .),

LAW OF UNITY

and placed beside another term, mean that it is combined with it, as one single term is with another:

Thus $A(B+C) = AB+AC$.

68. One plural term is combined with another by combining each alternative of the one separately with each of the other. Each combined alternative may then be combined with each alternative of a third plural term, and so on: *Combination of plural terms.*

Thus $(D+E)(B+C) = B(D+E) + C(D+E)$
$= BD+BE+CD+CE$.

69. It is in the nature of thought and things that *same alternatives are together same in meaning, as any one taken singly*. *Law of unity.*

Thus, what is the same as A *or* A is the same as A, a self-evident truth.

$$A+A = A \qquad A+A+A = A \qquad A+A+B = A+B$$

This law is correlative to the Law of Simplicity (§ 39), and is perhaps of equal importance and frequent use. It was not recognised by Professor Boole, when laying down the principles of his system.

70. In a plural term, any alternative may be removed, of which a part forms another alternative. *Superfluous terms.*

Thus the term *either B or BC* is the same in meaning with B alone, or $B+BC = B$. For it is a self-evident truth (§ 99) that B standing alone is either the same as BC, or as *B not-C*. Thus

$$B+BC = B \text{ } not\text{-}C + BC + BC$$
$$= B \text{ } not\text{-}C + BC = B.$$

71. A plural term obeys the Law of Simplicity (§ 42). For let $A = B+C$; then— *Plural terms obey laws of single terms.*

$AA = (B+C)(B+C)$.
$AA = BB+BC+BC+CC$ (§ 68).
$A = B+BC+C$ (§ 42).
$A = B+C$ (§ 70).

A plural term obeys the Law of Unity (§ 69):

$$A+A = B+C+B+C = B+C.$$

Substitution in plural terms.

72. For any alternative or part of an alternative may be substituted (§ 51) its expression in other terms:

Thus, if $A = B + CD$ and $D = E$, substitute, getting $A = B + CE$.

Substitution of plural terms.

73. A plural term may be substituted like a single term for any term, single or plural, of which it is the expression. When in combination, the several alternatives must be separately combined (§§ 66, 68).

Conversely, for a plural term may be substituted its expression in a single term:

Thus, if $A = BC$ and $C = D+E$, for C substitute $D+E$, and $A = B(D+E) = BD+BE$.

Or from the premises $A = BD + BE = B(D + E)$ and $C = D+E$, we might by substitution get back to $A = BC$.

Plural term known

74. *A plural term is known when each of its alternatives is known.*

Thus, in $A = B+C$, A is known when the meanings of each of B and C are known. But of course from knowing a single meaning of A, we cannot learn either or both of B and C.

Number of terms and premises.

75. With reference to the relation between the number of premises, and the numbers of known and unknown terms (§§ 33-35), we must treat as a separate term each alternative of a plural term.

A proposition with a plural term thus corresponds to an equation with several unknown quantities.

Plural and single terms.

76. As plural terms obey the laws of single terms, and a term single in form may be plural in meaning, it will not be necessary for the future to distinguish *plural and single terms*, any more than it has been to distinguish *combined and simple terms.*

There is some danger of misconception concerning plural terms. Though a plural term has one of several meanings, it cannot bear in this system more than one at the same time, so to speak. Hence it still remains a *unit*, the name of a single set of qualities, one of several sets, but it is not known which. The whole of this system in short is *unitary*, and involves the same remarkable analogies to a calculus of unity and 0 which have been brought forward so explicitly in Professor Boole's system.

CHAPTER VII

OF NEGATIVE PROPOSITIONS

TERMS may also be known and stated as differing, or not being the same in meaning.

Law of difference.
77. It is in the nature of thought and things that *a thing which differs from another differs from everything the same as that other.*

More briefly stated—*Same as different are different.*

Hence in logic—

A term which differs from another term in meaning differs from every term which is the same as that other.

If A is not the same as B, which is the same as C, then A is not the same as C. *The inference arises in the sameness of B and C, allowing us to substitute one for the other.* Hence we learn nothing of the sameness or difference of any two terms, D and E, each of which differs from a third, F; for D and E may each have any of an indefinite variety of meanings, and each may yet differ from F (§ 152).

Negative inference.
78. Hence a chain of related premises between any of which inferences can be drawn, must not contain more than a single negative premise. Also any inference in which a negative premise is concerned must be a negative inference.

Conversion.
79. *A negative proposition is simply convertible.* For *A is not the same as B*, is the same statement as *B is not the same as A*.

Law of different parts and wholes.
80. *When same terms are combined with different terms, the wholes may be different.*

If A differs from B, then AC differs from BC, provided, however, that the difference of A and B does not consist in any part of C.

81. *When from different wholes same parts are taken, the remainders are different.* — Law of different wholes.

This is equally self-evident with the preceding converse.

It is unnecessary further to consider negative propositions, because their inferences may be obtained by use of affirmative propositions.

CHAPTER VIII

OF CONTRARY TERMS

Negative term. **82.** *The known meaning of a negative term is the absence of the quality, or set of qualities, which forms the known meaning of a certain other, its positive term.*

Thus *not-A* is the negative term signifying the absence of the quality or set of qualities A. If the known meaning of A be only a single quality, *not-A* means its absence; but if A mean several qualities, *not-A* means the absence of any one or more.

Thus, if $A = B.C$

$$\textit{not-}A = B \;\textit{not-}C + \textit{not-}B.C + \textit{not-}B \;\textit{not-}C.$$

Negative of negative. **83.** *The negative of a negative term is the corresponding positive term.*

What is *not-not-A* is A.

Simple contrary terms defined. **84.** Since the relation of a positive to a negative term is the same as the relation of a negative to a positive, let each be called the *simple contrary term* of the other.

Notation. **85.** For convenience let *not-A* be written a. Then any large and its small letter denote a pair of simple contraries; and *not-a* is A. Also, the contrary of BC (§ 82) is

$$Bc + bC + bc,$$

which expresses the absence of one or more of B and C.

Laws obeyed. **86.** All that has been said of a term applies samely to one as to the other of a pair of contraries.

Thus, a obeys the several laws:

$$aa = a \quad a+a = a \quad \begin{matrix}C = D \\ aC = aD\end{matrix} \quad \text{and so on.}$$

87. Let a combined term or a proposition be said to *involve* a term when it contains either that term or its contrary. — *'Involve' defined.*

88. The contrary of a plural term is a term containing a contrary of each alternative. — *Contrary of plural term.*

Thus the contrary of $A+B+C$ is abc. If any alternative has more than one contrary, for each there will be a contrary alternative. Thus, $A+BC$ has the plural contrary $aBc+abC+abc$.

89. Any combined term which contains the simple contrary of another term may be called *a contrary*, or *contrary combination* of this, or of any combination containing this. — *Contrary combinations.*

Thus, any combined term containing A is a contrary of any term containing a, and it will seldom be necessary to distinguish by name *simple contraries*, such as A and a from contraries, or contrary combinations in general, which merely contain A or a. (See, however, §§ 99, 100.)

90. It is in the nature of thought and things that *a thing cannot both have and not have the same quality*. — *Law of contradiction.*

91. Hence a term which means a collection of qualities in which the same quality both *is and is not*, cannot mean the qualities of anything which is or ever will be known. — *Contradictory term defined.*

Such a term then has *no meaning*, that is to say, no possible, useful, or thinkable meaning; but it may be said to mean *nothing*. Let it be called a *self-contradictory*, or, for sake of brevity, a *contradictory term*.

92. Let us denote by the term or mark 0, combined with any term, that this is contradictory, and thus excluded from thought. Then $Aa = Aa.0$, $Bb = Bb.0$, and so on. For brevity we may write $Aa = 0$, $Bb = 0$. Such propositions are tacit premises of all reasoning. — *Use of 0.*

Any two contrary terms in combination give a contradictory term.

Combina-tion with contra-dictory.

93. Any term being combined with a contradictory, the whole is contradictory.

For the whole then means a collection of qualities which does and does not contain some same quality, and is therefore by definition a contradictory.

$$\text{Thus, if } A = Bb\ \ = Bb.0$$
$$AC = BbC = BbC.0.$$

Term 0.

94. The term 0, meaning *excluded from thought*, obeys the laws of terms.

$$0.0 = 0 \quad 0 + 0 = 0,$$

otherwise expressed :—What is excluded *and* excluded is excluded—What is excluded *or* excluded is excluded.

Condition of non-contradiction.

95. *Any term not known to be contradictory must be taken as not contradictory.*

Any term known to be contradictory is excluded from notice, and any term concerning which we are desiring knowledge must therefore be assumed not contradictory.

Contradictory alternatives.

96. *In a plural term of which not all the alternatives are contradictory, the contradictory alternative or alternatives must be excluded from notice.*

If for instance $A = 0 + B$, we may infer $A = B$, because A if it be 0 is excluded; and if it be such as we can desire knowledge of, it must be the other alternative B.

Elimination of contradictory.

97. *No contradictory term is to be eliminated in direct inference.*

For all we can require to know of a contradictory term is that it is contradictory, and elimination of a contradictory term would prevent rather than give such knowledge.

Thus if $A = Cc.0$, $B = Cc.0$, all that we can require to know of A and B is known from these premises, and cannot be known from the inference $A = B$ got by eliminating the contradictory $Cc.0$.

So, if $A = B = C = D = E = F = Gg.0$, the only useful inferences are those showing each of A, B, C, D, E, F, to be contradictory.

So, also, obviously, of intrinsic elimination.

It may be said, in fact, that contradiction supersedes all other elimination by itself eliminating all contradictory terms from further notice.

98. An alternative is eliminated when its plural term is combined with a contrary of that alternative. *Elimination of alternatives.*

Thus, the alternative Ab is removed from the plural term $AB+Ab$ when combined with B.

$$(AB+Ab)B = AB+ABb = AB+0 = AB$$
Let $\quad C = AB+Ab+aB+ab$
Then $\quad AC = AB+Ab \qquad ABC = AB$
$\qquad\quad BC = AB+aB \qquad AbC = Ab$
$\qquad\quad aC = aB + ab \qquad aBC = aB$
$\qquad\quad bC = Ab + ab \qquad abC = ab.$

The term thus combined with each side cannot be eliminated intrinsically (§ 53), and remains *a condition of the rejection of the other alternative*.

It is by this rejection of alternatives that the extent or width of the meaning of a term is reduced, as its intent of known meaning is increased, by combination (§ 1). For every general term, in addition to its known meaning, may be assumed to have an indefinite multitude of unknown alternatives. In combination with a new term many of these will probably become contradictory.

CHAPTER IX

OF CONTRARY ALTERNATIVES

Law of Duality. **99.** It is in the nature of thought and things that *a thing is either the same or not the same as another thing.* Otherwise—

A set of qualities either does or does not contain a certain quality.

Hence, in logic, a term must contain the meaning of one of any pair of *simple contrary terms.* Thus:—

A term is not altered in meaning by combination with any simple contrary terms as alternatives.

$$A = A(B+b) = AB + Ab.$$

For if A has meanings containing only B, then Ab is contradictory, and $A = AB + 0 = AB$.

If A has meanings containing only b, then $AB = 0$ and $A = 0 + Ab = Ab$.

If A has meanings of which some contain B and some b, the law is still true.

This Law of Duality is not the same as Professor Boole's law of duality. (See § 42.)

Apparent exceptions. **100.** Some apparent exceptions may occur to this law. For instance, let A = virtue, B = black, and b = not-black. Then the statement

Virtue is either black or not-black, seems true according to the above law, and yet absurd. This arises from B and b not being simple contraries; for B may be decomposed into *black-coloured*—say BC, and *b* into *not-black-coloured*, or *not black and not coloured*, or $bC+bc$. Now, virtue is really not coloured at all, or is Abc, and, therefore, neither BC nor bC. Here, again, we must observe that the combination Bc is contradictory from the tacit premise *black is a colour* (§ 48).

Other apparent inconsistencies may be similarly explained.

Professor De Morgan has excellently said,[1] ' It is not for human reason to say what are the simple attributes into which an attribute may be decomposed.' And for such a reason it is that I have as far as possible abstained from treating any term as *known to be simple*.

101. Let a term, combined with simple contraries as alternatives, be called a *development* of the term as regards the contraries. *Development defined.*

Thus, AB+Ab is called a development of A as regards B, or in terms of B, or involving B.

102. *Any term is same in meaning after combination with all the possible combinations of other terms, and their contraries as alternatives.* *Continued development.*

Since A = AB+Ab, and, again, A = AC+Ac, we may substitute for A in AB+Ab (§ 51) its expression in terms of C. Thus,

$$A = ABC + ABc + AbC + Abc.$$

Again, since A = AD+Ad, we may substitute a second time, getting

A = ABCD+ABCd+ . . . . . +Abcd, and so on.

103. Let any two alternatives, differing only by a single part-term and its contrary, be called a *dual term*. *Dual term defined.*

Thus, AB + Ab is a dual term as regards B, and ABC+ABc as regards C, and we may speak of B+b or C+c as the *dual part*.

104. *A dual term may always be reduced to a single term*

[1] *Syllabus*, p. 60.

Reduction of dual term. by removal of the contrary terms, without altering the meaning.

For the term thus obtained is, by the Law of Duality, the same in meaning as the former dual term (§ 99).

Thus, from such a term as $AB + Ab$, we may always remove the dual part $B + b$, and the meaning of the term A will still be as before, since $A = AB + Ab$ is a self-evident (§ 99) truth always in our knowledge.

CHAPTER X

OF CONTRARY TERMS IN PROPOSITIONS

105. *From any affirmative premise we may infer a negative* Affirmative *proposition by changing any term on one side only into its* into nega- *contrary.* tive proposition.

From $A = B$ we have A not $= b$; for evidently B is not $= b$, and hence, by Law of Difference (§ 77), $A = B$ not $= b$, or A not $= b$.

From $AB = AC$, similarly, AB not $= Ac$.

106. *The two terms of a negative proposition are con-* Terms of *traries.* negative proposition.

For the two terms of a negative proposition are different in meaning. Hence there must be some quality or qualities in the meaning of one, and not in that of the other; thus, the combination of the two terms would mean both the absence and presence of a certain quality or qualities, and would be a contradictory. The two terms then are contrary (§ 89).

107. *A negative proposition may be changed into an* Negative *affirmative, of which one term is a term of the negative, and* into affir- *the other term this term combined with the contrary of the* proposi- *other term of the negative.* tion.

Thus, if A not $= B$, then $A = Ab$; or, again, $B = aB$.

For developing A in terms of B (§ 101), we have $A = AB + Ab$, but A and B being contraries (§ 106), AB is contradictory or 0. Hence, $A = 0 + Ab = Ab$ (§ 96).

Similarly, we may show $B = 0 + aB = aB$. So, if AB

not = AC, then AB = ABc. For AB = ABC+ABc = ABc, since ABC is contradictory. And we see that

$$ABc = AB \text{ (contrary of AC)}$$
$$= AB\ (Ac+aC+ac) = ABc+0+0.$$

Since we may now convert any negative proposition into an affirmative, it will not be further necessary to use negative propositions in the process of inference (§ 81).

Inference of negative propositions.

108. *From any contradictory combination we may infer that any part of the combination not itself contradictory is not the same in meaning as the remainder or any greater part.* That the two parts differ may be expressed in a negative proposition, or its corresponding affirmative.

For if the other part be contradictory, it cannot be the same as the first part, which is not contradictory. And if neither of the parts is contradictory in itself, they cannot be same in meaning, else their combination would not produce a contradiction.

The affirmative inferences corresponding (§ 107) to the negative ones deduced under this rule may be otherwise had, so that it seems unnecessary to consider the negative inferences further in this place.

List of laws.

109. The following are the chief laws or conditions of logic:—

Condition or postulate. The meaning of a term must be same throughout any piece of reasoning; so that $A = A$, $B = B$, and so on. (§ 14.)

Law of Sameness. (§ 25.)

$$A = B = C;\ \text{hence}\ A = C.$$

Law of Simplicity. (§ 42.)

$$AA = A,\ BBB = B,\ \text{and so on.}$$

Law of Same Parts and Wholes. (§ 44.)

$$A = B;\ \text{hence}\ AC = BC.$$

Law of Unity. (§ 69.)

$$A+A = A,\ B+B+B = B,\ \text{and so on.}$$

Law of Contradiction. (§ 90.)

$Aa = 0$, $ABb = 0$, and so on.

Law of Duality. (§ 99.)

$A = A (B+b) = AB + Ab$
$A = A (B+b) (C+c)$
$\quad = ABC + ABc + AbC + Abc$, and so on.

It seems likely that these are the primary and sufficient laws of thought, and others only corollaries of them. Logic may treat only of *known samenesses of things;* and differences of things need be noticed, only for the exclusion from pure logical thought of all that is self-contradictory.

In pure number and its science, on the other hand, differences of things only are noticed.

The Laws of Simplicity, Unity, Contradiction, and Duality furnish the universal premises of reasoning. The Law of Sameness is of altogether a higher order, involving Inference, or the Judgment of Judgments.

CHAPTER XI

OF INDIRECT INFERENCE

Use of de-velopment. **110.** Taken by itself, the development of a term (§ 101) gives us no new knowledge about it. But taken with the premises of a problem, we may learn that some of the alternatives of the development are contradictory and to be rejected. The remaining alternatives then form a new and often useful expression for the term.

Indirect inference. **111.** In thus using a development we are said to infer *indirectly*, because we use the premise to show what a term is, not directly by the Law of Sameness, but *indirectly by showing what it is not*.

Indirect Inference is direct inference with the aid of self-evident premises derived from the Laws of Contradiction and Duality. But all Inference is still by the Law of Sameness.

Example. **112.** Let $A = B$: required expressions for A, B, a, b, inferred from this premise. Develop these terms as follows (§ 101):—

$$A = AB + Ab \qquad B = AB + aB$$
$$a = aB + ab \qquad b = Ab + ab$$

Examine which of the alternatives AB, Ab, aB, ab, are contradictory according to the premise $A = B$.

A combined with $A = B$ gives	$A = AB$
B „ „ „ „	$AB = B$
a „ „ „ „	$Aa = aB = 0$
b „ „ „ „	$Ab = Bb = 0$

Hence we learn that aB and Ab are contradictory, and may be rejected, and that AB is not contradictory according to the premise. Of ab, which is not found among any of the above terms, we can learn nothing from the premise, and it therefore cannot be known to be contradictory. Striking out aB and Ab in the developments of A, B, a, b, we have—

$$A = AB+0 = AB \qquad B = AB+0 = AB$$
$$a = 0 + ab = ab \qquad b = 0 + ab = ab$$

113. We have here the two inferences $A = AB \ B = AB$ *Inferences.* which might have been had from the premise by combination (§ 45), and from which we may pass back by elimination of AB to the premise.

We also have $a = ab$, and $b = ab$, *which could not have been had by direct inference.* And by eliminating ab between these two we have the new inference $a = b$. This result, indeed, that *from the sameness of meaning of two terms, we may infer the sameness of meaning of their simple contraries* is evidently true.

114. By a similar method we may draw inferences from *Inference from many* any number of premises, namely, by developing any re- *premises.* quired term in respect of other terms, and striking out the combinations which are shown to be contradictory in any premise.

Thus, from $A = B$ and $B = C$, to infer expressions for A and a, we develop these terms as follows :—

$$A = ABC + AB\mathit{c} + A\mathit{b}C + A\mathit{bc}$$
$$a = a\mathrm{BC} + a\mathrm{B}\mathit{c} + a\mathit{b}\mathrm{C} + a\mathit{bc}$$

By combination we then, when possible, render one side of each premise same with each of the alternative combinations, and learn from the other side whether the combination is known to be contradictory by the premise. All the combinations in the above developments will be found contradictory, except ABC and abc, and we thus get the inferences $A = ABC$, and $a = abc$, of which the former indeed might have been got directly.

PURE LOGIC

Method of indirect inference.
115. The process of indirect inference may similarly be applied to drawing any possible inference or expression from any series of premises, however numerous and complicated. The full process may be abbreviated according to the following series of rules, which may be said to form THE METHOD OF INDIRECT INFERENCE:—

Development.
1. Any premises being given, form a combination containing every term involved therein (§ 87). Change successively each simple term of this into its contrary, so as to form all the possible combinations of the simple terms and their contraries.

Comparison.
Included subject.
Excluded.
Contradiction.
2. Combine successively each such combination with both members of a premise. When the combination forms a contradiction with neither side of a premise, call it an *included subject* of the premise; when it forms a contradiction with both sides, call it an *excluded subject* of the premise; when it forms a contradiction with one side only, call it a *contradictory combination* or *subject*, and strike it out.

Possible.
Impossible.
We may call either an included or excluded subject a *possible subject*, as distinguished from a contradictory combination or *impossible subject*.

Repeated comparison.
3. Perform the same process with each premise. Then a combination is an included subject of a series of premises, when it is an included subject of any one; it is a contradictory subject when it is a contradictory of any one; it is an excluded subject when it is an excluded subject of *every* premise.

Selection.
4. The expression for any term involved in the premises consists of all the included and excluded subjects containing the term, treated as alternatives.

Reduction.
5. Such expression may be simplified by reducing all dual terms (§ 104), and by intrinsic elimination (§ 52) of all terms not required in the expression.

Elimination.
6. When it is observed that the expression of a term contains a combination which would not occur in the expression of any contrary of that term, we may eliminate the part of the combination common to the term and its expression. (See below, § 117.)

INDIRECT INFERENCE

7. Unless each term of the premises and the contrary of each appear in one or other of the possible subjects, the premises must be deemed inconsistent or contradictory. Hence there must always remain at least two possible subjects (§ 159). *Contradictory premises.*

116. Required by the above process the inferences of the premise $A = BC$. *Example.*

The possible combinations of the terms A, B, C, and their contraries, are as given in the margin. Each of these being combined with both sides of the premise, we have the following results:— *Development. Comparison.*

ABC	= ABC		ABC included subject	ABC
ABc	= ABCc = 0		ABc contradiction	ABc
AbC	= ABbC = 0		AbC contradiction	AbC
Abc	= ABbCc = 0		Abc contradiction	Abc
$0 = Aa$BC	= aBC		aBC contradiction	aBC
$0 = AaBc$	= aBCc = 0		aBc excluded subject	aBc
$0 = Aab$C	= aBbC = 0		abC excluded subject	abC
$0 = Aabc$	= aBbCc = 0		abc excluded subject	abc

It appears, then, that the four combinations ABc to aBC are to be struck out, and only the rest retained as possible subjects.

Suppose we now require an expression for the term b as inferred from the premise $A = BC$. Select from the included and excluded subjects such as contain b, namely abC and abc. *Selection.*

Then $b = ab$C $+ abc$, but as aC occurs only with b, and not with B, its contrary, we may, by Rule 6, eliminate b from abC; hence $b = a$C $+ abc$. *Elimination.*

117. The validity of this last elimination is seen by drawing the expression for aC, which is abC. Then between $b = ab$C $+ abc$, and abC $= a$C, we may eliminate abC by substituting (§ 51) its expression aC. And similarly in all other cases to which the rule applies. *Elimination explained.*

We might also reduce the expression for b by Rule 5, as follows:—

$$b = ab\text{C} + abc = ab\ (\text{C} + c) = ab.$$

Other inference.

118. To express a we have
$$a = aBc + abC + abc,$$
but observing that none of Bc, bC, bc, occur with A, so that $Bc = aBc$, $bC = abC$, $bc = abc$, we substitute these simpler terms, eliminating a; whence $a = Bc + bC + bc$, an evident truth (§ 113).

Other inferences.

119. Similarly, we may draw any of the following inferences:—

$$A = ABC = AB = AC$$
$$B = AC + aBc$$
$$C = AB + abC$$
$$c = aB + abc = ac$$
$$aB = Bc$$
$$aC = bC$$
$$ab = abC + abc = ab \text{ (no inference)}$$
$$ac = aBc + abc = ac \text{ (no inference)}.$$

Relation of B and C.

120. Observe that since B and C are samely related to A, we may get any inference concerning one of these terms from the similar inference concerning the other by interchanging B and C, b and c (§ 56).

Before proceeding to further examples of indirect inference, we may make the following observations.

Excluded subjects.

121. When any term appears on both sides of a premise, as A in $AB = AC$, any combination containing its contrary, a, is an excluded subject. Thus, in combining any term with both sides of a proposition, we render any contrary of the term an excluded subject.

So, in mathematics we introduce a new root into an equation when we multiply both sides by a factor.

Of inferior importance.

122. An excluded subject, though admitting of inference and admitted into inferences, is of inferior and often of no importance. As its name expresses, it is usually a combination concerning which we do not desire knowledge. The sphere of an argument, or the *Universe of Thought*, contains all the included subjects. An excluded subject is such as lies beyond this sphere or universe. But we are obliged to consider excluded subjects, because the excluded subject of one premise may be the included subject of other premises.

123. When a premise is plural in one or both sides, an excluded subject is a contrary of all the alternatives on both sides, and a contradictory combination is a contrary of all on one side, and not of all on the other side. *Plural premises.*

124. Of an identical proposition the term itself appearing on either side is the only included subject. All others are excluded, and there are no contradictory combinations. Its useless nature is thus evident. *Identical proposition.*

125. Any subject of a proposition remains an included, excluded, or contradictory subject as before, after combination with any unrelated terms. Thus, if the argument be restricted to a sphere or *common subject*, defined by certain terms, these do not need expression in each premise, but may be retained as an exterior condition. Thus, by ABCD (. . . . .) we might mean that ABCD is to be understood as combined with each term of any premises placed within the brackets. ABCD is then the common subject of the premises, which must contain no contrary of this. And any contrary of ABCD is an excluded subject of the whole. *Common subject.*

126. Any set of terms which always occur in the premises in unbroken combination may be treated as a simple term. *Of unbroken combinations.*

Thus, if BC occur always thus in combination, we may write for it, say D, and then d or *not*-BC is $bC + Bc + bc$.

127. Any set of alternatives which always occur together in the premises as alternatives may be treated as a single term. *Unbroken plural terms.*

Thus, if B and C occur always as alternatives, we may for B + C write, say D, and then d or *neither B nor C* is bc.

128. Any proposition may be treated under the form of

Simple proposition.

A = B, so long as we do not require to treat its part-terms or alternatives separately. (By §§ 126, 127.)

Technical terms.

129. Hence the convenience in every branch of knowledge of using technical terms to stand for every large set of terms which usually occur together. But such terms become the source of error if we do not carefully keep before us their definitions, those adopted premises in which we express the set of combined or alternative terms for which we substitute a technical term.

Metaphysical terms.

130. In that branch of knowledge, however, called First Philosophy, which is *analytic*, and aims at resolving things, or our thoughts about them, into their simplest components, the use of technical terms is fallacious. Such terms cannot assist analysis, since each arises from the synthesis of many simpler terms, forming its definition. All reasoning, then, in Metaphysics or First Philosophy, ought to be carried on in the simplest and most vernacular elements of speech. Analytic science should be like a mill which grinds down the ordinary grains of thought into their smallest and simplest particles. It is in the bakehouse we should combine these particles again into loaves of a size and consistency suitable for ordinary use. But most metaphysical reasoners, it seems to me, have mistaken the mill and the bakehouse.

Interrupted process.

131. It is not always necessary to carry out the process of inference exactly as in the rules. Each or any premise may be treated as a separate one, if desirable, and its possible subjects afterwards combined with the possible subjects of other premises. We may thus successively add premises, or try the effect of supposed ones.

For instance, since AB and *ab* are the possible subjects of A = B, and BC and *bc* of B = C, the possible combinations of these, namely ABC and *abc*, are the possible subjects of the two premises combined, observing that AB*bc* and *ab*BC are contradictory.

Unrelated premises.

132. If premises be related, the indirect inferences will include all possible direct inferences. From unrelated premises we shall also get such inferences as are possible.

Thus, from the unrelated premises

$$A = B, \ C = D,$$

we have

$$A = BCD + Bcd$$
$$d = ABc + abc, \text{ and so on.}$$

133. It does not seem possible to give any general proof that the conclusions of the indirect method must agree with those of the direct method, which will make its truth any the more evident. Such proof could be little less than a general recapitulation of the several Laws of Thought. *Proof of indirect method.*

134. It hardly needs to be pointed out that the method of indirect inference is equivalent to Euclid's indirect demonstration, or *reductio ad absurdum*. Euclid assumes the development of alternatives, usually that of *equal or greater or less*, and showing that two of these lead to a contradiction, establishes the truth of the third. *Euclid's indirect demonstration.*

135. Nor is this process of reasoning at all new or uncommon in any branch of knowledge save logic, which was supposed to be the science of all reasoning. Simple instances occur perhaps as frequently as instances of direct inference, and complicated instances are only rendered scarce by the limited powers of human memory and attention. Among instances of indirect argument we may place all those discourses in which a writer or speaker states several possible alternatives or cases of his subject, and, after showing some of them to be impossible, concludes the rest to be necessary, or else proceeds further to develop and consider these with regard to other premises (§ 131). A good instance is found in Paley's Argument on the Divine Benevolence (*Moral Phil.*, Book II, chap. v). The old logical process called *abscissio infiniti* has a close relation to indirect inference. *Common use of indirect method.*

136. Even brute animals, it would seem, may reason by the indirect method :— *Quotation.*

'This creature, saith Chrysippus (of the dog), is not void

of Logick: for, when in following any beast he cometh to three several ways, he smelleth to the one, and then to the second; and if he find that the beast which he pursueth be not fled one of these two ways, he presently, without smelling any further to it, taketh the third way; which, saith the same Philosopher, is as if he reasoned thus: the Beast must be gone either this, or this, or the other way; but neither this nor this; *Ergo*, the third: so away he runneth.'

Sir W. Raleigh.

CHAPTER XII

OF RELATION TO COMMON LOGIC

BEFORE giving examples of the processes of logical inference as now set forth, it will be well to consider the relation of our system to the logic of common thought.

137. In ordinary reasoning it will be found that there is *Ordinary* great economy of thought. Not only are large collections *proposition imperfect.* of attributes and things grouped together under the fewest possible terms, but only those particular attributes of the things under consideration on which the reasoning turns are brought forward. A certain natural disinclination to exertion causes us to simplify our modes of thought as much as possible, and to leave in the background everything that is not essential. Thus when we say *man is mortal,* we mean that the attributes of mortality are among the attributes of man. But we leave out those infinitely numerous attributes of man which are not comprised under mortality, because we do not happen to be occupied with them. The proposition, then, in this form is not that equation of qualities, that statement of perfect sameness or equivalence of meaning, which we have taken as a proposition.

138. It may be objected that we ought to take the pro- *Equation* position as we find it in common thought. Aristotle so *or sameness the* took it, and his system has had a long reign. Some of the *true form* expounders of his system even denied that there could be a *of reasoning.* proposition of two universal and equivalent terms. They could not have committed a greater error or more completely misrepresented the ordinary course of reasoning. Not only, as a fact, do the several sciences establish multitudes of propositions of which the two terms are equivalent and universal, but all definitions are propositions of this kind, and the definitions requisite in connecting the meanings of more and less complex terms, must always form a large part of our data in reasoning. If we further consider that even Aristotle's negative propositions have a universal predicate, that men show a constant tendency to treat the predicate of the proposition A as universal, whence several common kinds of fallacy, and that *reasoning from same to same things may be detected as the fundamental principle of all the sciences*, we need have no hesitation in treating the equation as the true proposition, and Aristotle's form as an imperfect proposition.

It is thus the Law of Sameness, not the *dictum* of Aristotle, which governs reasoning.

Quantification of predicate. **139.** It is only of very late years that the imperfection of the ordinary proposition has been properly pointed out. It is the discovery of the so-called *quantification of the predicate* which has reduced the proposition to the form of a convertible equation, and opened out to logic an indefinite field of improvement.

Boole's 'Analysis.' **140.** Professor Boole's system, first published in his *Mathematical Analysis of Logic*, in 1847, involves this newly discovered quantification of the predicate. According to Boole, the *some*, which is the adjective of particular logical quality, is an *indefinite class symbol*. *Men are some mortals* is expressed by him in the equation, $x = vy$, where x instructs us to select from the universe all things that are men, and y to select all things that are mortal. The proposition then informs us that the things which are men consist of an indefinite selection from among the things which are mortal, v being the symbol of this indefinite quantity or class selected.

Further step requisite. **141.** One more step seems to me necessary. It is to separate completely the qualitative and quantitative meanings of all logical terms, including the word *some*. In the qualitative form of the proposition *man is some mortal*—or more correctly speaking, *man is some kind of mortal*—we interpret *some* or *some kind* as meaning an indefinite and perhaps unknown collection of qualities, which being added to the qualities *mortal*, give the known qualities of *man*. In the quantitative form *men are some mortals*, we have the equivalent statement that the collection of individuals in the class *some mortals* is the collection of individuals in the class *men*.

142. It is strange that the purely qualitative form of proposition *man is some kind of mortal*, which is the most distinct form of statement, and is perhaps the most prevalent, both in science and ordinary thought, was totally disregarded by logicians, at least as the foundation of a system of logic. 'The Logicians, until our day,' says Professor De Morgan,[1] 'have considered the extent of a term as the only object of logic, under the name of the *logical whole*: the *intent* was called by them the *metaphysical whole*, and was excluded from logic.' *Qualitative proposition disregarded.*

143. It will be seen that this word *some* or *some kind*, the source of so much difficulty and error, must in our system be treated as a term of indefinite and unknown meaning. It is an unknown term, not only at the beginning of a problem, but throughout it. In no two premises then can the term *some* or *some kind* be taken to mean the same set of qualities. Thus we cannot argue through or eliminate a term with *some*, while at least it retains this unknown term: that is to say, we can never use it as a common term (§ 27) in direct inference. Thus, if A is *some* B, and *some* B is *some* C, we cannot eliminate *some* B getting A *is some* C, because *some* being of unknown meaning, the *some* B is not necessarily the same in both cases. This is still more plain in the form *A is some kind of B*, and *some kind of B is some kind of C*, for it is obvious that the one kind of B is not necessarily the same as the other. *'Some' and 'some kind.'*

144. Since the term *some* or *some kind* is not only unknown but remains unknown throughout any argument, we *Symbol for 'some.'*

[1] *Syllabus*, p. 61.

might conveniently appropriate to it some symbol such as U, to remind us of its special conditions. Thus no term U is to be taken as same with any other term U, or $U = U$ is not known to be true. But in the propositions A and E it is always open to us, and is best to eliminate U by writing for it the other member of the proposition (§ 52). Thus, $A = UB$, meaning that A is *some kind* of B, involves three terms. It is much better written as $A = AB$, involving only A and B, and yet perfectly expressing that the qualities of B are among those of A, but not necessarily those of A all among those of B.

Aristotle's propositions.

145. The four propositions of the old logic may thus find expression in our system:—

A	*Every A is B*	$A = UB$ or	$A = AB$
E	*No A is B*	$A = Ub$ or	$A = Ab$
I	*Some A is B*	$UA = UB$ or	$CA = DB$
O	*Some A is not B*	$UA = Ub$ or	$CA = Db$.

De Morgan's propositions.

146. Two new propositions of De Morgan's system are thus expressed:—

Everything is either A or B	$A = b$
Some things are neither A nor B	$a = b$.

Thomson's proposition.

147. All these propositions, and as many more as may be proposed, can be brought and partially treated (§ 128) under the form $A = B$, which I believe to be the simple form of all reasoning. The existence of doubly-universal propositions of this kind was far from being unknown to many of the School Logicians, but out of deference to the Aristotelian system, such propositions were neglected. The present Archbishop of York first embodied this proposition in a system of logic, giving it the name U. (Thomson's *Outlines, passim*).

CHAPTER XIII

EXAMPLES OF THE METHOD

In this chapter I shall place some miscellaneous examples of inference according to the system of the foregoing chapters, suited to show the power of its method, or its relation to the old logic.

148. Let us take a syllogism in FELAPTON. *Syllogism in Felapton.*

 No A is B $\quad A = Ub = Ab$
 Every A is C $\quad A = UC = AC$
 Some C is not B $\quad$ (§ 145).

From $A = Ab$ we might by combination (§ 45) infer *Direct inference.* $AC = AbC$, and from $A = AC$, $Ab = AbC$; whence $AC = AbC = Ab$, or $AC = Ab$, *which is a more precise statement of* $UC = Ub$, or *some* C *is not* B, the Aristotelian conclusion.

We may, however, obtain this conclusion, as well as all other possible ones, by indirect inference.

Of the possible combinations of A, B, C, a, b, c, ABC and ABc are contradicted by the first premise, and Abc (as well as ABc) is contradicted by the second premise. AbC, aBC, aBc, abC, abc, are the remaining combinations in which we find there is no relation between B and C *per se*, since B occurs with C and c, and C occurs with B and b. But $AC = AbC$, and $Ab = AbC$, whence, by elimination, $AC = Ab$, the same conclusion as before.

AbC
aBC
aBc
abC
abc

The following conclusions may also be drawn:

PURE LOGIC

$$a = a(BC + Bc + bC + bc) = a(B+b)(C+c)$$
$$= a \text{ (no inference)}$$

$$B = aB$$
$$b = AC + abC + abc = AC + ab$$
$$C = Ab + aBC + abC = Ab + aC$$
$$c = aBc + abc = ac$$
$$ab = abC + abc = ab(C+c) = ab \text{ (no inference)}$$
$$aC = aBC + abC = aC(B+b) = aC \text{ (no inference)}$$
$$bc = abc.$$

Example:
AB = CD

ABCD
A*b*C*d*
A*bc*D
A*bcd*
*a*BC*d*
*a*B*c*D
*a*B*cd*
*ab*C*d*
*abc*D
abcd

149. The premise AB = CD is of some interest. It contradicts the combinations ABC*d*, AB*c*D, AB*cd*, which are AB and not CD, and A*b*CD, *a*BCD, *ab*CD, which are CD and not AB. From the remainder we easily draw the inferences.

$$A = BCD + AbCd + AbcD + Abcd$$
$$a = BCd + BcD + Bcd + abCd + abcD + abcd.$$

Observing that A and B enter samely into the premise, we may easily deduce the expressions for B and *b* by interchanging A and B, *a* and *b* in the above; thus (§§ 54-56)—

$$B = ACD + aBCd + aBcD + aBcd.$$

And since A, B enter samely with C, D, we might deduce the corresponding expressions for C, D, and *c*, *a*, by interchanging at once A with C, and B with D, or A with D, and B with C.

From the expression for *a* we thus get

$$d = CBa + CbA + Cba + dcBa + dcbA + dcba.$$

Observe, that if the expression for A be combined with that for *a*, nothing but contradictory terms will be the result, verifying A*a* = 0. And, if we combine the expressions for any terms not contrary, as B and *d*, we get the same result as we might have drawn by the separate application of the process.

Thus, B*d* = A*b*CD + *a*BC*d* + *a*B*cd* = 0 + *a*B*d*.

EXAMPLES

In expressions thus derived there will often appear, as in the above instance, a superfluous and contradictory term (AbCD, a contrary of Bd), but being only an alternative, the proposition is not untrue.

150. As an example of a premise with a plural term, let us take A = B+C.

Example:
A = B + C

In comparing the eight combinations of A, B, C, a, b, c, with the premise, any one is contradictory which contains A without containing either B or C; or, again, which contains either B or C without containing A. Thus, ABC, ABc, and AbC, are the included subjects, abc is an excluded subject, and the rest are contradictory.

ABC
ABc
AbC
abc

We may draw the inferences

$$A = BC + Bc + bC$$
$$a = bc$$
$$B = ABC + Ac = AB$$
$$b = AbC + a$$
$$C = ABC + Ab = AC$$
$$c = ABc + a$$

Observe that B and C enter samely, so that their expressions may be mutually derived by interchange.

151. The premise A = Bc+bC differs from the last in the very important point that A cannot at once be B and C.

Example:
A = Bc
  + bC

It has the included subjects ABc and AbC, and the excluded subjects aBC and abc. The following expressions are seen to be simple and symmetrical, and it is instructive to form their combinations.

ABc
AbC
aBC
abc

$$A = Bc + bC$$
$$a = BC + bc$$
$$B = Ac + aC$$
$$b = AC + ac$$
$$C = Ab + aB$$
$$c = AB + ab.$$

152. From two negative premises we can infer no Aristotelian conclusion (§§ 77, 78). It is well to show that this

Negative premises.

remains true when the negative propositions are converted into their corresponding affirmatives (§ 107).

Let us take the premises

> A is not the same as B,
> C is not the same as B.

These may be expressed by the affirmative propositions $A = Ab$, $C = bC$.

AbC
Abc
aBc
abC
abc

If we go through the process of indirect inference, and attempt to express A and C in terms of each other, we shall obtain:—

$$A = AbC + Abc = Ab(C+c) = Ab$$
$$C = AbC + abC = bC(A+a) = bC.$$

These are the premises over again, and there can be no new inference, except $B = aBc$.

Solutions of form $A = B$

153. The proposition $A = B$ being the simplest form of statement, its full solution is given below, and the solutions of the similar propositions $A = b$, $a = B$, $a = b$, are inferred by interchanging A and a, B and b.

Premise	$A = B$	$A = b$	$a = B$	$a = b$
Included subject	AB	Ab	aB	ab
Excluded subject	ab	aB	Ab	AB
Contradiction	Ab	AB	AB	aB
Contradiction	aB	ab	ab	Ab.

Example of two premises.

154. Let us take $A = ABC$,
$$B + C = BD + CD.$$

We have, by direct inference from the second premise,

$$BC = BCD \quad (\S 45)$$

Hence $\quad A = ABC = ABCD \quad (\S 26).$

ABCD
*a*BCD
*a*BcD
*ab*CD
*abc*D
abcd

The indirect process gives four included and two excluded subjects, as in the margin.

Hence not only the above inference, but the following, among other possible ones:—

$$a = a\mathrm{BCD} + a\mathrm{B}c\mathrm{D} + ab\mathrm{CD} + abc\mathrm{D} + abcd$$
$$ = a\mathrm{BD} + ab\mathrm{D} + abd$$
$$ = a\mathrm{D} + abd = a\mathrm{BD} + ab$$
$$\mathrm{C} = \mathrm{ABD} + a\mathrm{BCD} + ab\mathrm{CD} = \mathrm{ABD} + a\mathrm{CD}$$
$$c = a\mathrm{B}c\mathrm{D} + abc\mathrm{D} + abcd = ac\mathrm{D} + abd$$
$$\mathrm{B}c = a\mathrm{B}c\mathrm{D} \qquad\qquad c\mathrm{D} = ac\mathrm{D}$$

155. The ordinary Sorites is easily and clearly solved in this system. Taking four premises such as *Sorites.*

$\mathrm{A} = \mathrm{AB}$

$\mathrm{B} = \mathrm{BC}$

$\mathrm{C} = \mathrm{CD}$

$\mathrm{D} = \mathrm{DE}$, many inferences will be evident from the following series of the subjects, or possible combinations.

$$\left.\begin{array}{l}\mathrm{ABCDE}\\ a\mathrm{BCDE}\\ ab\mathrm{CDE}\\ abc\mathrm{DE}\end{array}\right\} \text{Included subjects.}$$

$$\left.\begin{array}{l}abcd\mathrm{E}\\ abcde\end{array}\right\} \text{Excluded subjects.}$$

156. The Dilemma of the old logic is easily included in our system, when we supply a term which is suppressed or understood in its usual statement. The dilemma is as follows:— *Dilemma.*

If A is B, E is F, and if C is D, E is F: but, either A is B, or C is D, therefore E is F. Adopting Wallis's reduction to the categorical form, and supplying some term G, to express the *present circumstances*, or the *case* in which either A is B, or C is D, we have the premises

$$\mathrm{AB} = \mathrm{ABEF}$$
$$\mathrm{CD} = \mathrm{CDEF}$$
$$\mathrm{G} = \mathrm{ABG} + \mathrm{CDG}.$$

By the direct process alone we get the required conclusion that, under the condition G, E is F; thus—

$$\mathrm{GE} = (\mathrm{AB} + \mathrm{CD})\ \mathrm{GE} = \mathrm{ABEFG} + \mathrm{CDEFG} = \mathrm{GEF}$$
$$\text{or, } \mathrm{GE} = \mathrm{GEF}.$$

Destructive Conditional Syllogism.

157. The following is known as a Destructive Conditional Syllogism.

If A is B, C is D; but C is not D; therefore, A is not B.

Supplying the suppressed term, say E, expressing the circumstances in which A is not B, the following is the statement of this syllogism in our system:—

$$AB = ABCD$$
$$CE = CdE.$$

By direct inference

$$ABE = ABD.CE = ABD.CdE = 0.$$

Hence ABE is known to be contradictory; therefore (§ 108), AE is not ABE, or in the circumstances E, A is not B.

Forms of old logic.

158. The forms of the old logic being comprehended in this system along with an indefinite multitude of other forms, logicians can only properly accept this generalisation, due to Boole, by throwing off as dead encumbrances the useless distinctions of the Aristotelian system. The past history of the Science must not, as hitherto, bar its progress. And Logic will be developed almost like Mathematics, when Logicians like Mathematicians discriminate between the Study of Thought and the Study of Antiquarian Lore.

I will now give a few complex problems, more suited to show the power of the method.

Complex problem.

159. Let the premises be

$$A = B + C$$
$$B = c + d$$
$$c = cD$$
$$AD = BCD.$$

And let it be required to infer the description of any term, say a. By the indirect process, we shall find that the only combination uncontradicted by one or other premise is ABCd.

Thus, we find there cannot be any a at all, without

EXAMPLES

contradiction, whatever may be the meaning of this result.[1] It means, doubtless, that the premises are contradictory (§ 115.7).

We also easily infer any of the following:—

$$A = BCd \qquad AB = Cd \qquad ABC = ABCd$$
$$B = ACd \qquad AC = Bd \qquad ABd = ABCd$$
$$C = ABd \qquad BC = Ad \qquad \text{etc.}$$
$$d = ABC.$$

160. The following premises are such as might easily occur in physical science:— *Problem.*

$$A = ABc + AbC$$
$$B = BDE + Bdc$$
$$C = CDe.$$

---

[1] The following law, being of a less evident character than the rest, has been placed apart. *Law of infinity.*

Every logical term must have its contrary.

That is to say:—*Whatever quality we treat as present we may also treat as absent.*

There is thus no boundary to the universe of logic. No term can be proposed wide enough to cover its whole sphere; for the contrary of any term must add a sphere of indefinite magnitude. Let U be the universe; then u is not included in U. Nor will special terms limit the universe. *Universe of logic unbounded.*

Thing existing has its contrary in *thing not existing.*

Thing thinkable has its contrary in *thing not thinkable.*

Even *thing*, the widest noun in the language, has a contrary in *that which is not a thing.*

Of course the above is only true speaking in the strictest logical sense, and using all terms in the most perfect generality.

If the above be granted as true, every proposition of the form $A = B + b$ must be regarded as contradictory of a law of thought. For the contrary of A from the above is Bb, a contradiction, or A is used as having no contrary, and forming the universe. *Contradictory proposition.*

Also every system of premises must be rejected which altogether contradicts any term or terms. Thus in the indirect process we must always have at least two combinations remaining possible, one of which must contain the contrary of each simple term in the other. In this view the peculiar premises *Contradictory premises.*

$$A = B + C$$
$$B = c + d$$

(§ 159) contain subtle contradictions. For a according to the first premise must be bc, and being c, it must by the second premise be B, and hence by first premise also, A, or both A and a, B and b.

But this subject needs more consideration.

60 PURE LOGIC

ABcDE
ABcde
AbCDe
aBcDE
aBcde
abCDe
abcDE
abcDe
abcdE
abcde

The series of possible combinations in the margin gives by inspection perhaps the most useful information, but the following are a few formal inferences.

$$A = ABcDE + ABcde + AbCDe$$
$$BcD = BcDE$$
$$abd = abcd$$
$$cd = ABcde + aBcde + abcd(E+e)$$
$$= Bcde + abd$$
$$bCD = AbCDe + abCDe$$
$$= bCDe.$$

There is no relation between abc and D and E.

Complicated problem. **161.** I conclude with the solution of a still more complicated system of premises.

$$A + C + E = B + D + F$$
$$Bc + bC = Dc + dE$$
$$AD = ADf$$
$$D = c$$
$$C = Cd.$$

The possible combinations are:—

ABcDcf   aBcdEF
ABcdEF   aBcdEf
ABcdEf   abCdEF
AbCdEF.

Whence the following, among many other inferences, may be drawn:—

$$A = BcDef + ABcdE + AbCdEF$$
$$Bc = ADef + BcdE$$
$$D = ABccf$$
$$cd = BE$$
$$c = ABcDf$$
$$dE = ABcd + bCF + aBc = Bcd + bCF$$
$$bdE = bCF = CdE$$
$$AF = ABcdEF + AbCdEF$$
$$aF = aBcdEF + abCdEF$$

$$df = \text{B}cd\text{E}f$$
$$b = b\text{C}d\text{EF}$$
$$\text{C} = b\text{C}d\text{EF} = b.$$

Whence the remarkable and unexpected relation $\text{C} = b$, which it would not be easy to detect in the premises.

162. Inferences may be verified by combining the expressions of two or more terms, and comparing the result with the expression of the combined term as drawn from the series of possible combinations. For instance, in the problem last given (§ 161), we may combine the expression for A with that for dE, as follows :— *Verification.*

$$\text{A}.d\text{E} = (\text{B}c\text{D}ef + \text{AB}cd\text{E} + \text{A}b\text{C}d\text{EF})(\text{B}cd + b\text{CF})$$
$$= 0 + \text{AB}cd\text{E} + 0 + 0 + \text{A}b\text{C}d\text{EF},$$

the contradictory combinations being struck out. But the expressions thus obtained may not always be in their simplest terms.

163. The reduction of inferences to their simplest terms, it may be remarked, is in no way essential to their truth; it only renders them more pregnant with information. It is, perhaps, the only part of the process in which there is any difficulty. *Reduction not necessary.*

164. In working these logical problems, it has been found very convenient to have a series of combinations of terms beginning with those of A, B, and proceeding up to those of A, B, C, D, E, F, or more, *engraved* upon a common writing slate. In any given problem, the series is chosen which just furnishes sufficient letters for the distinct terms. The contradictory combinations may then be rapidly struck out, and the remaining combinations lie ready before the eye. *Working of the process.*

CHAPTER XIV

COMPARISON WITH BOOLE'S SYSTEM

165. To show the power and facility of this method, as compared with that of Professor Boole, it will be sufficient, as regards those already acquainted with Professor Boole's system, to present the solution of one of his complex examples. Thus, let us follow Professor Boole's investigation of Senior's definition of wealth, namely[1]—that *wealth is what is transferable, limited in supply, and either productive of pleasure or preventive of pain* (Boole, p. 106).

Senior's definition of wealth.

Let A = Wealth
    B = Transferable
    C = Limited in supply
    D = Productive of pleasure
    E = Preventive of pain.

The definition in question is expressed by the proposition

$$A = BC(DE + De + dE)$$

which includes all the combinations of D, E, d, e, except de.

Striking out the dual term $(E + e)$ from BCD $(E + e)$, we may state the definition in the more concise form

$$A = BCD + BCdE.$$

We may pass over Professor Boole's expression for A, after

---

[1] Here, as usually elsewhere, I take words in intent of meaning, and transform most statements accordingly.

intrinsic elimination of E (A = BCD+ABCd), as being sufficiently obvious.

166. *Required* C *in terms of* A, B, D (Boole, p. 107). Forming all the possible combinations of A, B, C, D, E, and their contraries, and comparing them with the premise, we shall find all the combinations from ABCde to aBCdE inclusive contradicted. The remaining subjects are as in the margin.

Expression for C.
ABCDE
ABCDe
ABCdE
...
...
aBCde
aBcDE
aBcDe
aBcdE
aBcde
abCDE
abCDe
abCdE
abCde
abcDE
abcDe
$abcd$E
$abcde$

Selecting the terms containing C, we have

C = ABCDE+ABCDe+ABCdE+aBCde
  +abCDE +abCDe +abCdE +abCde.

Striking out the dual terms (E+e), and intrinsically eliminating remaining E's or e's by substitution of C, we have

C = ABCD+ABCd+aBCd+abCD+abCd.

Eliminating C from ABCD (§ 117), because ABD = ABCD, and striking out the dual terms (A+a) and (D+d), we have either of the expressions—

C = ABD+BCd +abC
C = ABC +aBCd+abC.

From the latter we read, *What is limited in supply is either wealth, transferable* (*and either productive of pleasure or not,* ABC), *or else some kind of what is not wealth, but is either not transferable* (abC), *or, if transferable, is not productive of pleasure* (aBCd).

This conclusion is exactly equivalent to that of Professor Boole, on p. 108.

167. His so-called secondary propositions, namely, ' 1. Wealth that is intransferable and productive of pleasure, does not exist;' and ' 2. Wealth that is intransferable and not productive of pleasure does not exist,' are negative conclusions implied in the striking out of the contradictory combinations AbCDE, AbCDe, A$b$$cDE, Ab$$cDe$, and A$bCdE, AbCde$, A$bcdE, Abcde$, which are easily reducible to

AbD (C+c) (E+e) = 0   AbD = 0
Abd (C+c) (E+e) = 0   Abd = 0.

Negative conclusions.

The expression 'does not exist' is open to exception.

Expression for D.
168. Again, required an expression for productive of pleasure (D), in terms of wealth (A), and preventive of pain (E) (Boole, p. 111).

The complete collection of combinations containing D is

ABCDE	abCDE
ABCDe	abCDe
aBcDE	abcDE
aBcDe	abcDe.

We may then write D as follows:—

$$D = ABCDE + ABCDe + a(Bc + bC + bc)(E + e)D.$$

But we may observe also that

$$ADE = ABCDE \text{ and } Ae = ABCDe.$$

Hence we may substitute ADE and Ae for the two first terms of the expression for D. We may also strike out the dual term (E+e) in the third term, and eliminate the plural term (Bc+bC+bc) intrinsically (§ 52) by substitution of D. Thus we get the expression in the required terms:

$$D = ADE + Ae + aD,$$

which may be translated into these words:—*What is productive of happiness is either some kind of wealth preventive of pain, or any kind of wealth not preventive of pain, or some kind of what is not wealth* (Boole, p. 111 *ad fin.*)

Expression for d.
169. For the expression of d we easily select

$$d = AdE + ade + adE = AdE + ad$$

of which the meaning is—*What is not productive of pleasure is either some kind of wealth preventive of pain, or some kind of what is not wealth* (Boole, p. 112).

Other inferences.
170. These are the chief inferences furnished by Mr. Boole. From the list of possible combinations we could easily add a great many more inferences, in fact as many more, as may be drawn concerning any of the five terms A, B, C, D, E, and their contraries.

Thus for CE expressed in the remaining terms, we have

$$CE = ABCDE + ABCdE + abCDE + abCdE$$
$$= (ABCE + abCE)(D+d).$$

Striking out the dual term $(D+d)$ and extrinsically eliminating C in ABCE, since we observe that ABE = ABCE, we have

$$CE = ABE + abCE$$

which may be translated—

What is limited in supply, and preventive of pain, is either wealth, transferable and preventive of pain, or some kind of what is not wealth and not transferable.

But we may often find that there is no special relation to express. Thus, in trying to express $abCD$ in terms of E we find

$$abCD = abCD(E+e) = abCD.$$

171. Besides affording these formal deductions, the series of possible combinations will often give us at a glance a clear and valuable notion of the manner in which the universe of our subject is made up. *General character of combinations.*

In this instance we see that for wealth we have the three combinations BCDE, BCDe, and BCdE, and that thus for *not-wealth* (a) we have all possible combinations of B, C, D, E, except those three. With aB we have Cde and c (DE+De+dE+de), and with ab, we have all possible combinations of C, D, and E. Thus the definition gives no relation between what is not wealth and not transferable, and what is limited in supply, productive of pleasure, or preventive of pain.

172. It is the character of this logical system, in common with that of Professor Boole, that it is perfectly general. The same rules which govern the inferences from one or two premises, involving two or three terms, are applicable without the slightest modification to any number of premises, involving any number of terms. Of course the working of the inferences becomes rapidly more laborious as the com- *Generality of the system.*

plexity of the problem increases, and a considerable liability to mistake arises. But this is in the nature of things, and the process of inference, consisting in the mere comparison of terms as to their sameness or difference, seems to me the simplest process that can be conceived.

Comparison with Boole's system.

173. Compared with Professor Boole's system, in its mathematical dress, this system shows the following advantages :—

1. Every process is of self-evident nature and force, and governed by laws as simple and primary as those of Euclid's axioms.

2. The process is infallible, and gives no uninterpretable or anomalous results.

3. The inferences may be drawn with far less labour than in Professor Boole's system, which generally requires a separate computation and development for each inference.

CHAPTER XV

REMARKS ON BOOLE'S SYSTEM, AND ON THE RELATION OF LOGIC AND MATHEMATICS

174. So long as Professor Boole's system of mathematical logic was capable of giving results beyond the power of any other system, it had in this fact an impregnable stronghold. Those who were not prepared to draw the same inferences in some other manner could not quarrel with the manner of Professor Boole. But if it be true that the system of the foregoing chapters is of equal power with Professor Boole's system, the case is altered. There are now two systems of notation, giving the same formal results, one of which gives them with self-evident force and meaning, the other by dark and symbolic processes. The burden of proof is shifted, and it must be for the author or supporters of the dark system to show that it is in some way superior to the evident system.

175. It is not to be denied that Boole's system is consistent and perfect within itself. It is, perhaps, one of the most marvellous and admirable pieces of reasoning ever put together. Indeed, if Professor Ferrier, in his *Institutes of Metaphysics*, is right in holding that the chief excellence of a system is in being *reasoned* and consistent within itself, then Professor Boole's is nearly or quite the most perfect system ever struck out by a single writer.

176. But a system perfect within itself may not be a perfect representation of the natural system of human thought. The laws and conditions of thought as laid down in the system may not correspond to the laws and conditions of thought in reality. If so, the system will not be one of Pure and Natural Logic. Such is, I believe, the case. Professor Boole's system is Pure Logic fettered with a condition which converts it from a purely logical into a numerical system. His inferences are not logical inferences; hence they require to be *interpreted*, or translated back into logical inferences, which might have been had without ever quitting the self-evident processes of pure logic.

Among various objections which I might urge to Boole's system, *regarded as purely logical in purpose*, are four chief ones to which I shall here confine my attention.

First Objection

177. *Boole's symbols are essentially different from the names or symbols of common discourse—his logic is not the logic of common thought.*

Professor Boole uses the symbol + to join terms together, on the understanding that they are logical contraries, which cannot be predicated of the same thing or combined together without contradiction. He says (p. 32)—' In strictness, the words "and," "or," interposed between the terms descriptive of two or more classes of objects, imply that those classes are quite distinct, so that no member of one is found in another.'

178. This I altogether dispute. In the ordinary use of these conjunctions, we do not necessarily join logical contraries only ; and when terms so joined do prove to be logically contrary, it is by virtue of a *tacit premise*, something in the meaning of the names and our knowledge of them, which teaches us they are contrary. And when our knowledge of the meanings of the words joined is defective, it will often be impossible to decide whether terms joined by conjunctions are contrary or not.

179. Take, for instance, the proposition—' A peer is either a duke, or a marquis, or an earl, or a viscount, or a baron.' If expressed in Professor Boole's symbols, it would be implied that a peer cannot be at once a duke and marquis, or marquis and earl. Yet many peers do possess two or more titles, and the Prince of Wales is Duke of Cornwall, Earl of Chester, Baron Renfrew, etc. If it were enacted by parliament that no peer should have more than one title, this would be the tacit premise which Professor Boole assumes to exist.

Again,—' Academic graduates are either bachelors, masters, or doctors,' does not imply that a graduate can be only one of these ; the higher degree does not annul the lower.

Shakespeare's lines—

> 'Beauty, truth, and rarity,
> Grace in all simplicity,
> Here inclosed in cinders lie.
>
> . . . .
>
> To this urn let those repair
> That are either true or fair,'—

certainly do not imply that beauty, truth, rarity, grace, and the true and fair are incompatible notions, so that no instance of one is an instance of another.

In the sentence—'Repentance is not a single act, but a habit or virtue,' it cannot be implied that a virtue is not a habit; by Aristotle's definition it is.

Milton has the expression in one of his sonnets—'Unstain'd by gold or fee,' where it is obvious that if the fee is not always gold, the gold is a fee or bribe.

Tennyson has the expression 'wreath or anadem.' Most readers would be quite uncertain whether a wreath may be an anadem, or an anadem a wreath, or whether they are quite distinct or quite the same.

From Darwin's *Origin*, I take the expression, 'When we see any *part or organ* developed in a remarkable *degree or manner*.' In this, *or* is used twice, and neither time disjunctively. For if *part* and *organ* are not synonymous, at any rate an organ is a part. And it is obvious that a part may be developed at the same time both in an extraordinary degree and manner, although such cases may be comparatively rare.

180. From a careful examination of ordinary writings, it will be found that the meanings of terms joined by 'and' 'or' vary from absolute identity up to absolute contrariety. There is no logical condition of contrariety at all, and when we do choose contrary expressions, it is because our subject demands it. The matter, not the form of an expression, points out whether terms are exclusive. And if there is one point on which logicians are agreed, it is that logic is formal, and pays no regard to anything not formally expressed. (See § 48.)

181. And if a further proof were wanted that Professor Boole's symbols do not correspond to those of language, we have only to turn to his own work. He actually translates one same sentence into different sets of symbols, according to the view he takes of the matter in hand. For instance (p. 59) he interprets 'Either productive of pleasure or preventive of pain' so as not to exclude things both productive of pleasure and preventive of pain. 'It is plain,' he remarks, 'from the nature of the subject, that the expression "either productive of pleasure or preventive of pain," in the above definition, is meant to be equivalent to "either productive of pleasure; or, if not productive of pleasure, preventive of pain."'

And in remarking upon other possible interpretations, he says, 'That before attempting to translate our data into the rigorous language of symbols, it is above all things necessary to ascertain the *intended* import of the words we are using' (p. 60). This simply amounts to consulting the matter, and Professor Boole's symbols thus constantly imply restrictions not expressed in the forms of language, but existing, if at all, as tacit or understood premises.

182. In my system, on the contrary, I take $A + B$ not to imply at all that A may not be B, but if this be the case, it must be owing to an expressed premise $A = Ab$ or $A = b$.

183. How essential Professor Boole's restriction on his symbols is to the stability of his system, one instance will show. Take his proposition on p. 35—

$$x = y + z,$$

and give the following meanings to x, y, z:—

x = Cæsar
y = Conqueror of the Gauls
z = First emperor of Rome.

Now, there is nothing logically absurd in saying 'Cæsar is the conqueror of the Gauls or the first emperor of Rome.'

It is quite conceivable that a person should remember just enough of history to make this statement and nothing more. And there is nothing in the logical character of the terms to decide whether the conqueror could or could not be the same person as the first emperor.

But now take Professor Boole's inference from the proposition $x = y + z$, namely $x - z = y$ got by subtracting z from either side of $x = y + z$. Then we have the strange inference:—

Cæsar, provided he is not the first emperor of Rome, is the conqueror of the Gauls.

This leads me to my second objection to Professor Boole's system.

Second Objection

184. *There are no such operations as addition and subtraction in pure logic.*

The operations of logic are the combination and separation of terms, or their meanings, corresponding to multiplication and division in mathematics. I cannot support this statement without going at once to the gist of the whole matter.

185. Number, then, and the science of number, arise out of logic, and the conditions of number are defined by logic. It has been thought that units are units inasmuch as they are perfectly similar. For instance, three apples are three units, inasmuch as each has exactly the same qualities as the other in being an apple. The truth is exactly opposite to this. *Units are units inasmuch as they are logically contrary.* In so far as three apples are exactly like each other, one could not be distinguished from the other. Were there three apples, or any three things, so perfectly similar in every way that we could not tell the difference, they would be but one thing, just as, by the law of unity before stated, $A + A + A = A$. But then we must remember that among the logical characters of a thing is its position in space with relation to other things, not to speak of its position in time. Now, when we speak of three apples, we mean three things, which, however perfectly same they may be in all other qualities, occupy different places, and are therefore distinct things. *In so far as they are same they are one; in that they are different they are three.*

186. The meaning of an abstract unit is *something only known as logically distinct from or contrary to other things.* The meaning of a concrete unit is *the abstract unit with certain qualities known or defined.*

For instance, in $A (1' + 1'' + 1''') = A' + A'' + A'''$ the meaning of the units $1', 1'', 1'''$, is that each is something logically distinct from the other, and when we predicate of each of these that it is A, say an apple, we get three distinct A's, $A' + A'' + A'''$.

So in multiplication, *twice two is four—*

$$(1 + 1)(1 + 1) = 1 + 1 + 1 + 1.$$

The logical significance of the process is that if we have two logically distinct notions, and we divide each into two logically distinct notions, we get four logically distinct notions. In logical formulæ $(A + a)(B + b) = AB + Ab + aB + ab$, where A and a, B and b, express logical contraries.

187. Now addition, subtraction, multiplication, and division, are alike true as modes of reasoning in numbers, where we have the logical condition of a unit as a constant restriction. But addition and subtraction do not exist, and do not give true results, in a system of pure logic, free from the condition of number.

For instance, take the logical proposition—

$$A + B + C = A + D + E$$

Meaning *what is either A or B or C is either A or D or E*, and *vice versâ*.

There being no exterior restrictions of meaning whatever, except that the same term must always have the same meaning (§ 14), we do not know which of A, D, E, is B, nor which is C; nor, conversely, do we know which of A, B, C, is D, nor which is E. The proposition alone gives us no such information.

In these circumstances, the action of subtraction does not apply. It is not necessarily true that, if from same (equal) things we take same (equal) things the remainders are same (equal). It is not allowable for us to subtract the same thing (A) from both sides of the above proposition, and thence infer—

$$B + C = D + E.$$

This is not true if, for instance, each of B and C is the same as E, and D is the same as A, which has been taken away.

Yet the equivalent inference by combination will be valid. We may combine a with both sides of the proposition, and we have

$$aA + aB + aC = aA + aD + aE$$

or, striking out the contradictory terms aA, we have

$$aB + aC = aD + aE.$$

188. But subtraction is valid under the logical restriction that the several alternatives of a term shall be mutually exclusive or contrary. Let

(1) $\quad$ $AMN + BMn + CmN = AMN + DMn + EmN$

in which it is obviously impossible that AMN can be either DMn or EmN, contraries of AMN, or any one of the three alternatives any other. Then we may freely subtract AMN from both sides, getting the necessary inference

(2) $\quad\quad\quad$ $BMn + CmN = DMn + EmN.$

This subtraction, however, is merely equivalent to the combination with both sides of the proposition (1) of the term (Mn + mN); for the combination being performed, and contradictory terms struck out, it will be found that the proposition (2) results.

189. In short, when alternatives are contraries of each other, subtraction of one is exactly equivalent to combination with the rest. The axiom (Boole, p. 36), that 'if equal things are taken from equal things, the remainders are equal,' is nothing but a case of the Law of Combination (§ 44), that if same (equal) terms be combined with same (equal) terms, the wholes are same (equal).

Take the self-evident proposition

$$AB + Ab + aB + ab = AB + Ab + aB + ab.$$

Any terms, say $aB + ab$, may be subtracted from both sides by combining the other terms $AB + Ab$ with each side of the proposition. Then

$$(AB + Ab)\ (AB + Ab + aB + ab) = (AB + Ab)\ (AB + Ab + aB + ab)$$
$$AB + Ab + 0\ ..\ + 0 = AB + Ab + 0 +\ ..\ 0 + 0.$$

And what is true of this self-evident case must be true when the premise is not self-evident.

190. Having thus established our liberty to subtract same terms, provided all alternatives are contraries, we have the corresponding liberty to add by the inverse process.

191. The processes of addition and subtraction thus arise out of the logical process of combination. The axioms of addition and subtraction are only valid under a logical condition, which is certainly not applicable to thought or language generally. And this condition is that which logic imposes upon number, that each two units shall be contrary logical alternatives. It is logic which reduces to a unit, by the Law of Unity, $A + A = A$, any two alternatives known to be the same, so that the science of number treating of units, treats of alternatives known to be different or contrary. But *logic itself is the superior science, and may treat of alternatives of which it is not known whether they are same or different.*

192. It is the self-evident logical Law of Unity, then, which lays the foundations of number. This law merely amounts to saying that a thing cannot and must not be distinguished from itself. We commit an error against this law, when in counting over coins, for instance, to ascertain their numbers, that is, how many logically distinct coins there are, we count the self-same coin two or more times, making the coins for instance

$$C' + C'' + C'' + C''' + C'''' + \ .\ .\ .\ .\ .\ .$$

instead of $C' + C'' + C''' + C'''' + \ldots \ldots$ It is by the Law of Unity that $C'' + C'' = C''$, or the same coin counted twice is but one coin in number. In this case no attention is paid to differences of time; but in many cases, things otherwise perfectly the same, like the beats of a pendulum, are distinguished and made into different units by one being before or after the other in time.

Third Objection

193. My third objection to Professor Boole's system is, that *it is inconsistent with the self-evident law of thought, the Law of Unity* $(A + A = A)$.

Professor Boole having assumed as a condition of his system that each two terms must be logically distinct, is unable to recognise the Law of Unity. It is contradictory of the basis of his system. The term x, in his system, means *all things with the quality x*, denoting the things in extent, while connoting the quality in intent. If by 1 we denote all things of every quality, and then subtract, as in numbers, all those things which have the quality x, the remainder must consist of all things of the quality *not-x*. Thus $x + (1 - x)$ means in his system *all x's with all not-x's*, which, taken together, must make up all things, or 1. But let us now attempt by multiplication with x, to select *all x's* from this expression for *all things*.

$$x(x + 1 - x) = x + x - x.$$

Professor Boole would here cross out one $+ x$ against one $- x$, leaving one $+ x$, the required expression for *all x's*. It is surely self-evident, however, that $x + x$ is equivalent to x alone, whether we regard it in extent of meaning, as *all the x's added to all the x's*, which is simply *all the x's*, or in intent of meaning, as *either x or x*, which is surely x. Thus, $x + x - x$ is really 0, and not x, the required result, and it is apparent that the process of subtraction in logic is inconsistent with the self-evident Law of Unity.

194. It is probable, indeed, that Professor Boole would altogether refuse to recognise such an expression as $x + x - x$, on the ground that it does not obey the condition of his symbols that each two alternatives shall be distinct and contrary, $x + x$ not being so. It may be answered, that the expression has been arrived at by operations enounced as universally valid, which ought to give true results. And if it be simply said that $x + x - x$ is not *interpretable* in Professor Boole's system, it may be again answered, that when translated into its equivalent in words, the expression $x + x - x$ has a very plain meaning. It is

REMARKS ON BOOLE'S SYSTEM

'*either x or x, provided it be not x*,' and this, I must hold, is simply *not x*, although it ought to be *x*, according to the mode in which it was got.

195. In founding his system, Boole assumed that there cannot be two terms A + B, the same in meaning or names of the same thing; the laws of thought require nothing of the kind, and cannot require it, because among known and unknown terms, any two such as A + B may prove to be names of the same thing AB. Thought merely reduces the meaning of two same terms AB + AB, by the Law of Unity, to be the same as that of one term AB. And when it is once known that all terms in question are contraries of each other, or naturally exclusive and distinct, then Boole's system and the whole science of numbers apply.

196. It is on this account that my objections have no bearing against Professor Boole's system as applied to the Calculus of Probabilities, so far as I can understand the subject. For it is a high advantage to that calculus to have to treat only events mutually exclusive, probabilities being then capable of simple addition and substraction.[1] It seems likely, indeed, that this distinction of exclusive and unexclusive alternatives is the Gordian knot in which all the abstract logical sciences meet and are entangled.

Fourth Objection

197. The last objection that I shall at present urge against Professor Boole's system is, that *the symbols $\frac{1}{1}$, $\frac{0}{0}$, $\frac{0}{1}$, $\frac{1}{0}$, establish for themselves no logical meaning, and only bear a meaning derived from some method of reasoning not contained in the symbolic system.* The meanings, in short, are those reached in the self-evident indirect method of the present work.

198. Professor Boole expressly allows, as regards one of these symbols at least, $\frac{0}{0}$, that it is not his method which gives any meaning to the symbol. It is the peculiarity of his system, that he bestows a meaning on his symbols by *interpretation*. The interpretation of $\frac{0}{0}$ is explained on pp. 89, 90, and he says, 'Although the above determination of the significance of the symbol $\frac{0}{0}$ is founded upon the examination of a particular case, yet the principle involved in the demonstration is general, and there are no circumstances under which the symbol can present itself to which the same mode of analysis is inapplicable.' Again (p. 91), 'Its actual interpretation, however, as an indefinite class symbol, cannot, I conceive, except upon the ground

[1] See De Morgan's *Syllabus*, § 243.

of analogy, be deduced from its arithmetical properties, but must be established experimentally.'

199. If I understand this aright, it simply means, that whereever a term appears in a conclusion with the symbol $\frac{0}{0}$ affixed, we may, by a mode of analysis, by some process of pure reasoning apart from the symbolic process from which $\frac{0}{0}$ emerged, ascertain that the meaning of $\frac{0}{0}$ is *some*, an *indefinite class term*. The symbol $\frac{0}{0}$ is unknown until we give it a meaning. Before, therefore, we can know what meaning to give, and be sure that this meaning is right, it seems to me we must have another distinct and intuitive system by which to get that meaning. Professor Boole's system, then, as regards the symbol $\frac{0}{0}$, is not the system bestowing certain knowledge; it is, at most, a system pointing out truths which, by another intuitive system of reasoning, we may know to be certainly true.

200. It is sufficient to show this with regard to a single symbol $\frac{0}{0}$, because the incapacity of a system, even in a single instance, proves the necessity for another system to support it. I believe that the other symbols, $\frac{1}{1}$, $\frac{1}{0}$, $\frac{0}{1}$, are open to exactly the same remarks, but from the way in which Mr. Boole treats them, involving the whole conditions of his system, it would be a lengthy matter to explain.

201. The obscure symbols $\frac{1}{1}$, $\frac{0}{0}$, $\frac{1}{0}$, $\frac{0}{1}$, have the following correspondence with the forms of the present system. $\frac{1}{1}$ appearing as the coefficient of a term means that the term is an *included subject* of the premise, so that, if combined with both members of the premise, it produces a self-contradictory term with neither side (§ 115).

Similarly, $\frac{0}{0}$ means that the term is an *excluded subject* of the premise, producing a self-contradictory with both sides of the premise.

And either $\frac{1}{0}$ or $\frac{0}{1}$ means that the term is a *self-contradictory or impossible term*, producing a self-contradictory term with one side only of the premise.

202. The correspondence of these obscure forms with the self-evident inferences of the present system is so close and obvious, as to suggest irresistibly that Professor Boole's operations with his abstract calculus of 1 and 0, are a mere counterpart of self-evident operations with the intelligible symbols of pure logic. Professor Boole starts from logical notions, and self-evident laws of thought; he suddenly transmutes his formulæ into obscure mathematical counterparts, and after various intricate manœuvres, arrives at certain forms, corresponding to forms arrived at directly and intuitively by ordinary or Pure Logic—by that analysis, from which the interpretation of his symbols was reached and proved. And by this interpretation he transfers the meaning and force of pure logical conclusions to obscure forms, which, if they have meaning, have certainly no demonstrative force of themselves. Boole's system is like the shadow, the ghost, the reflected image of logic, seen among the derivatives of logic.

203. Supposing it prove true that Professor Boole's Calculus of 1 and 0 has no real logical force and meaning, it cannot be denied that there is still something highly remarkable, something highly mysterious in the fact, that logical forms can be turned into numeral forms, and while treated as numbers, still possess formal logical truth. It proves that there is a certain identity of logical and numerical reasoning. Logic and mathematics are certainly not independent. And the clue to their connection seems to consist in distinct logical terms forming the units of mathematics.

204. Things as they appear to us in the reality of nature, are clothed in inexhaustible attributes, set as it were in a frame of time and space. By our mental powers we abstract first time, then space, and then attribute after attribute, until we can finally think of things as abstract units deprived of all attributes, and only retaining the original logical condition of things, that each is distinct from others. In logic we argue upon things as same and one, in number we reason upon them as distinct and many.

205. Supposing it be ultimately allowed that Professor Boole's calculus of 1 and 0 is not really logic at all; that his system is founded upon one condition, that of exclusive terms, which does not belong to thought in general, but only numerical thought; and that it ignored one law of logic, the Law of Unity, which really distinguishes a logical from a numerical system—these errors scarcely detract from the beauty and originality of the views he laid open. Logic, after his work, is to logic before his work, as mathematics with equations of any degree are to mathematics with equations of one or two degrees. He generalised logic so that it became possible to obtain any true inference from premises of any degree of complexity, and the work I have attempted has been little more than to translate his forms into processes of self-evident meaning and force.

Owens College, Manchester,
November 1863.

II

THE SUBSTITUTION OF SIMILARS,

THE

TRUE PRINCIPLE OF REASONING

THE LOGICAL ABACUS.

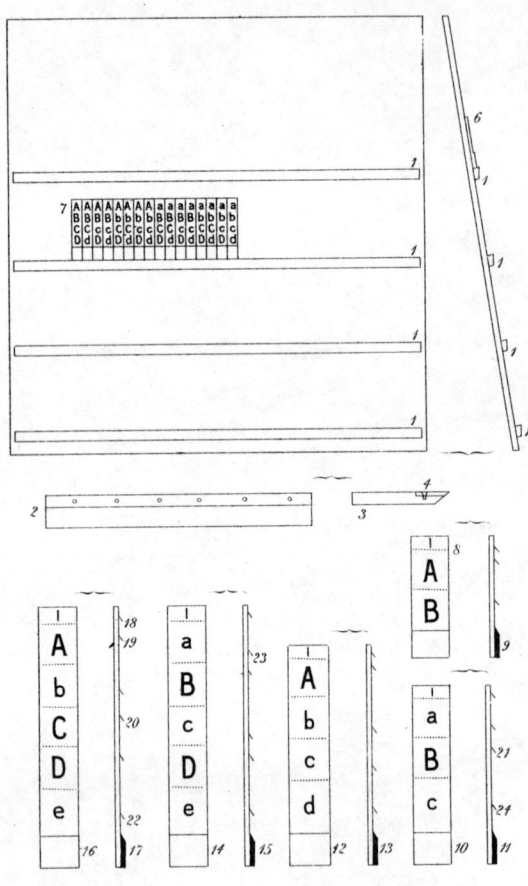

Plate 1

THE SUBSTITUTION OF SIMILARS

ARISTOTLE is, perhaps, the greatest of human authors, but we may apply to him the words of Bacon, 'Let great authors have their due, as Time, the author of authors, be not deprived of his due, which is farther and farther to discover truth.' Aristotle has had his due in the obedience of more than twenty centuries, and Time must not be deprived of his due. Men, whose birthright is the increasing result of reason, are not to be bound for ever by the *dictum* of a thinker who lived but a little after the dawn of scientific thought. We are not to be persuaded any longer to look upon the highest of the sciences as a dead science. Logic is the science of the laws of thought itself, and there is no sphere of observation and reflection which is more peculiarly open to any inquirer, than the inquirer's own mind as engaged in the process of reasoning. It is from reflection on the operations of his own mind that Aristotle must have drawn the materials of his memorable Analytics. But Bentham's mind, as he himself remarked, was equally open to Bentham,[1] and it would be slavery indeed if any *dictum* of the first of logicians were to deprive all his successors of the liberty of inquiry.

2. It may be said, perhaps, that the weaker cannot

[1] *Essay on Logic*, Bentham's works, vol. viii, p. 218.

possibly push beyond the stronger, and it is willingly allowed that among us moderns can few or none be found to equal in individual strength of intellect the great men of old. But Time is on our side. Though we reverence them as the ancients, they really lived in the childhood of the human race, and *these* times are, as Bacon would have said, the ancient times.[1] We enjoy not only the best intellectual riches of the Greeks and Romans, but also the wonderful additions to the physical and mathematical sciences made since the revival of letters. In our time we possess an almost complete comprehension of many parts of physical science which seemed to Socrates, the wisest of men, beyond the powers of the human mind. We have before us an abundance of examples of the modes in which solid and undoubted truths may be attained, and it is absurd to suppose that among such successful exertions of the human intellect we can find no materials for a newer analytic of the mental operations.

3. The mathematics especially present the example of a great branch of abstract science, evolved almost wholly from the mind itself, in which the Greeks indeed excelled, but in which modern knowledge passes almost infinitely beyond their highest efforts. Intellects so lofty and acute as those of Euclid or Diophantus or Archimedes reached but the few first steps on the way to the widening generalisations of modern mathematicians; and what reason is there to suppose that Aristotle, however great, should at a single bound

---

[1] 'De antiquitate autem, opinio quam homines de ipsa fovent, negligens omnino est, et vix verbo ipsi congrua. Mundi enim senium et grandævitas pro antiquitate vere habenda sunt; quæ temporibus nostris tribui debent, non juniori ætati mundi, qualis apud antiquos fuit. Illa enim ætas respectu nostri, antiqua et major; respectu mundi ipsius, nova et minor fuit. Atque revera quemadmodum majorem rerum humanarum notitiam, et maturius judicium, ab homine sene expectamus, quam a juvene, propter experientiam, et rerum quas vidit et audivit, et cogitavit, varietatem et copiam; eodem modo et a nostra ætate (si vires suas nosset, et experiri et intendere vellet) majora multo quam a priscis temporibus expectari par est; utpote ætate mundi grandiore, et infinitis experimentis, et observationibus aucta et cumulata.'—*Novum Organum*, Lib. i Aphor. 84.

have reached the highest generalisations of a closely kindred science of human thought ?

4. Kant indeed was no intellectual slave, and it might well seem discouraging to logical speculators that he considered logic unimproved in his day since the time of Aristotle, and indeed declared that it could not be improved except in perspicuity. But his opinions have not prevented the improvement of logical doctrine, and are now effectually disproved. A succession of eminent men,—Jeremy Bentham, George Bentham, Sir William Hamilton, Professor De Morgan, Archbishop Thomson, and the late Dr. Boole,—have shown that in the operations and the laws of thought there is a wide and fertile area of investigation. Bentham did more than assert our freedom of inquiry; in his uncouth logical writings are to be found most original hints, and in editing his papers his nephew George Bentham pointed out the all-important key to a thorough logical reform, the *quantification of the predicate*.[1] Sir William Hamilton, Archbishop Thomson, and Professor De Morgan, rediscovered and developed the same new idea. Dr. Boole, lastly, employing this fundamental idea as his starting-point, worked out a mathematical system of logical inference of extraordinary originality.

5. Of the logical system of Mr. Boole Professor De Morgan has said in his "Budget of Paradoxes":[2] 'I might legitimately have entered it among my *paradoxes*, or things counter to general opinion : but it is a paradox which, like that of Copernicus, excited admiration from its first appearance. That the symbolic processes of algebra, invented as tools of numerical calculation, should be competent to express every act of thought, and to furnish the grammar and dictionary of an all-containing system of logic, would not have been believed until it was proved. When Hobbes, in the time of the Commonwealth, published his *Computation or Logique*, he had a remote glimpse of some of the points

[1] See *Outline of a New System of Logic*, by George Bentham, Esq., London, 1827, p. 133 *et seq*.
[2] No. xxiii, *Athenæum*.

which are placed in the light of day by Mr. Boole. The unity of the forms of thought in all the applications of reason, however remotely separated, will one day be matter of notoriety and common wonder; and Boole's name will be remembered in connection with one of the most important steps towards the attainment of this knowledge.'

6. I need hardly name Mr. Mill, because he has expressly disputed the utility and even the truthfulness of the reforms which I am considering, and has evolved most divergent opinions of his own in a wholly different direction from the eminent men just mentioned.

7. In the lifetime of a generation still living the dull and ancient rule of authority has thus been shaken, and the immediate result is a perfect chaos of diverse and original speculations. Each logician has invented a logic of his own, so marked by peculiarities of his individual mind, and his customary studies, that no reader would at first suppose the same subject to be treated by all. Yet they treat of the same science, and, with the exception of Mr. Mill, they start from almost the same discovery in that science. Modern logic has thus become mystified by the diversity of views, and by the complication and profuseness of the formulæ invented by the different authors named. The quasi-mathematical methods of Dr. Boole especially are so mystical and abstruse, that they appear to pass beyond the comprehension and criticism of most other writers, and are calmly ignored. No inconsiderable part of a lifetime is indeed needed to master thoroughly the genius and tendency of all the recent English writings on Logic, and we can scarcely wonder that the plain and scanty outline of Aldrich, or the sensible but unoriginal elements of Whately, continue to be the guides of a logical student, while the works of De Morgan or of Boole are sealed books.

8. The nature of the great discovery alluded to, *the quantification of the predicate*, cannot be explained without introducing the technical terms of the science. A proposi-

tion, or judgment expressed in words, consists of a *predicate* or *attribute* united by a *copula* to a *subject*. In this proposition,

All metals are elements,

the predicate *element* is asserted of the subject *metal*, and the force of the assertion consists, as usually considered, in making the class of metals a part of the class of elements. The verb, or copula, *are*, denotes *inclusion* of the metals among the elements. But the subject only is quantified; for it is stated that *all metals* are elements, but it is not stated what proportion of the elements may be metals. Now the quantification of the predicate consists in giving some indication of the quantity or portion of the predicate really involved in the judgment.

All metals are some *elements*

is the same proposition thus quantified, and, though the change seems trifling, the consequences are momentous. The proposition no longer asserts the inclusion of one class in the other, but the identity of group with group. The proposition becomes *an equation* of subject and predicate, and the significance of this change will be fully apparent only to those who see that logical science thus acquires a point of contact with mathematical science. Nor is it only in a single point that the two great abstract sciences meet. Dr. Boole's remarkable investigations prove that, when once we view the proposition as an equation, all the deductions of the ancient doctrine of logic, and many more, may be arrived at by the processes of algebra. Logic is found to resemble a calculus in which there are only two numbers, 0 and 1, and the analogy of the calculus of quality or fact and the calculus of quantity proves to be perfect. Here, in all probability, we shall meet a new instance of the truth observed by Baden Powell, that all the greatest advances in science have arisen from combining branches of science hitherto distinct, and in showing the unity of principles pervading them.[1]

9. And yet any one acquainted with the systems of the modern logicians must feel that something is still wanting. So much diversity and obscurity are no usual marks of truth, and it is almost incredible that the true general system of inference should be beyond the comprehension of nearly every one, and therefore incapable of affecting ordinary thinkers. I am thus led to believe that the true clue to the analogy of mathematics and logic has not hitherto been seized, and I write this tract to submit to the reader's judgment whether or not I have been able to detect this clue.

10. During the last two or three years the thought has constantly forced itself upon my mind, that the modern logicians have altered the form of Aristotle's proposition without making any corresponding alteration in the *dictum* or self-evident principle which formed the fundamental postulate of his system. They have thus got the right form of the proposition, but not the right way of using it. Aristotle regarded the proposition as stating the inclusion of one term or class within another; and his axiom was perfectly adapted to this view.

The so-called *Dictum de omni* is, in Latin phrase, as follows—

Quicquid de omni valet, valet etiam de quibusdam et singulis.

And the corresponding *Dictum de nullo* is similarly—

Quicquid de nullo valet, nec de quibusdam nec de singulis valet.

In English these *dicta* are usually stated somewhat as follows—

Whatever is predicated affirmatively or negatively of a whole class may be predicated of anything contained in that class. Or, as Sir W. Hamilton more briefly expresses them, *What pertains to the higher class pertains also to the lower.*[2]

---

[1] Baden Powell, *Unity of the Sciences*, p. 41.
[2] *Lectures on Logic*, vol. i, p. 303.

THE TRUE PRINCIPLE OF REASONING 87

These *dicta*, then, enable us to pass from the predicate to the subject, and to affirm of the subject whatever we know or can affirm of the predicate. But we are not authorised to pass in the other direction, from the subject to the predicate, because the proposition states the inclusion of the subject in the predicate, and not of the predicate in the subject.

The proposition,

All metals are elements,

taken in connection with the *dictum de omni* authorises us to apply to *all metals* whatever knowledge we may have of the nature of *elements*, because metals are but a subordinate class included among the elements; and, therefore, possessing all the properties of elements. But we commit an obvious fallacy if we argue in the opposite direction, and infer of elements what we know only of metals. This is neither authorised by Aristotle's *dictum*, nor would it be in accordance with fact. Aristotle's postulate is thus perfectly adapted to his view of the nature of a proposition, and his system of the syllogism was admirably worked out in accordance with the same idea.

11. But recent reformers of logic have profoundly altered our view of the proposition. They teach us to regard it as an equation of two terms, formerly called the subject and predicate, but which, in becoming equal to each other, cease to be distinguishable as such, and become convertible. Should not logicians have altered, at the same time and in a corresponding manner, the postulate according to which the proposition is to be employed? Ought we not now to say that whatever is known of either term of the proposition is known and may be asserted of the other? Does not the *dictum*, in short, apply in both directions, now that the two terms are indifferently subject and predicate?

12. To illustrate this we may first quantify the predicate of our own former example, getting the proposition,

All metals are some *elements,*

where the copula *are* means no longer *are contained among,* but *are identical with;* or availing ourselves of the sign = in a meaning closely analogous to that which it bears in mathematics, we may express the proposition more clearly as,

All metals = some elements.

It is now evident that whatever we know of a certain indefinite part of the elements we know of all metals, and whatever we know of all metals we know of a certain indefinite part of the elements. We seem to have gained no advantage by the change; and if we are asked to define more exactly what part of the elements we are speaking of, we can only answer, *Those which are metals.* The formula

All metals = all metallic elements

is a more clear statement of the same proposition with the predicate quantified; for while it asserts an identity it implies the inclusion of metals among elements. But it is an accidental peculiarity of this form that the *dictum* only applies usefully in one direction, since if we already know what metals are we must know them to be *metallic* elements, the adjective *metallic* including in its meaning all that can be known of metals; and from knowing that metals are metallic elements we gain no clue as to what part of the properties of metals belong to elements. But it is hardly too much to say that Aristotle committed the greatest and most lamentable of all mistakes in the history of science when he took this kind of proposition as the true type of all propositions, and founded thereon his system. It was by a mere fallacy of accident that he was misled; but the fallacy once committed by a master-mind became so rooted in the minds of all succeeding logicians, by the influence of authority, that twenty centuries have thereby been rendered a blank in the history of logic.

THE TRUE PRINCIPLE OF REASONING

13. Aristotle ignored the existence of an infinite number of definitions and other propositions which do not share the peculiarity of the example we have taken. If we define elements as *substances which cannot be decomposed*,[1] this definition is of the form—

$$Elements = undecomposable\ substances\ ;$$

and since the term *element* does not occur in the second member, we may apply the *dictum* usefully in both directions. Whatever we know of the term *element* we may assert of the distinct term *undecomposable substance;* and, *vice versâ*, whatever we know of the term *undecomposable substance* we may assert of *element*.

The example,

Iron is the most useful of the metals;

hardly needs quantification of the predicate, for it is evidently of the form—

$$Iron = the\ most\ useful\ of\ the\ metals,$$

the terms being both singular terms, and convertible with each other. We may evidently infer of both terms what we know of either. If we join to the above the similar proposition,

$$Iron = the\ cheapest\ of\ the\ metals,$$

we are easily enabled to infer that the *cheapest of the metals* = *the most useful of the metals*, since by the *dictum* we know of iron that it is *the cheapest of the metals;* and this we are enabled to assert of *the most useful*, and *vice versâ*. These are almost self-evident forms of reasoning, and yet they were neither the foundation of Aristotle's system, nor were they included in the superstructure of that system. His syllogism was therefore an edifice in which the corner-stone

[1] In strictness we should add, *by our present means*.

itself was omitted, and the true system is to be created by supplying this omission, and re-erecting the edifice from the very foundation.

14. I am thus led to take the equation as the fundamental form of reasoning, and to modify Aristotle's *dictum* in accordance therewith. It may then be formulated somewhat as follows—

Whatever is known of a term may be stated of its equal or equivalent.

Or in other words,

Whatever is true of a thing is true of its like.

I must beg of the reader not to prejudge the value of this very evident axiom. It is derived from Aristotle's *dictum* by omitting the distinction of the subject and predicate; and it may seem to have become thereby even a more transparent truism than the original, which has been condemned as such by Mr. J. S. Mill and some others. But the value of the formula must be judged by its results; and I do not hesitate to assert that it not only brings into harmony all the branches of logical doctrine, but that it unites them in close analogy to the corresponding parts of mathematical method. All acts of mathematical reasoning may, I believe, be considered but as applications of a corresponding axiom of quantity; and the force of the axiom may best be illustrated in the first place by looking at it in its mathematical aspect.

15. The axiom indeed with which Euclid begins to build presents at first sight little or no resemblance to the modified *dictum*. The axiom asserts that

Things equal to the same thing are equal to each other.

In symbols,

$$a = b = c$$
gives $a = c$.

Here two equations are apparently necessary in order that an inference may be evolved; and there is something

THE TRUE PRINCIPLE OF REASONING

peculiar about the threefold symmetrical character of the formula which attracts the attention, and prevents the true nature of the process of mind from being discovered. We get hold of the true secret by considering that an inference is equally possible by the use of a single equation, but that when there is no equation no inference at all can be drawn. Thus if we use the sign $\backsim$ to denote the existence of an inequality or difference, then one equality and one inequality, as in

$$a = b \backsim c,$$

enable us to infer an *inequality*

$$a \backsim c.$$

Two inequalities, on the other hand, as in

$$a \backsim b \backsim c,$$

do not enable us to make any inference concerning the relation of a and c; for if these quantities are equal, they may both differ from b, and so they may if they are unequal. The axiom of Euclid thus requires to be supplemented by two other axioms, which can only be expressed in somewhat awkward language, as follows—

If the first of three things be equal to the second, but the second be unequal to the third, the first is unequal to the third.

And again—

If two things be both unequal to a third common thing, they may or may not be equal to each other.

16. Reflection upon the force of these axioms and their relations to each other will show, I think, that the deductive power always resides in an equality, and that difference as such is incapable of affording any inference. My meaning will be more plainly exhibited by placing the symbols in the following form:—

$$a = b \qquad a$$
$$\| \quad \text{hence} \quad \|$$
$$c \qquad\qquad c.$$

Here the inference is seen to be obtained by substituting a for b by virtue of their equality as expressed in the first equation $a = b$, the second equation $b = c$ being that in which substitution is effected. One equation is *active* and the other is *passive*, and it is a pure accident of this form of inference that either equation may be indifferently chosen as the active one. Precisely the same result happens in this case to be obtained by a similar act of reasoning in which $b = c$ is the active equation, as shown below—

$$b = c \qquad c$$
$$\| \quad \text{hence} \quad \|$$
$$a \qquad\qquad a.$$

My warrant for this view of the matter is to be found in the fact that the negative form of the axiom is now easily brought into complete harmony with the affirmative form, except that, since it has only one equation to work by, there can be only one active equation and one form in which the inference can be exhibited as below—

$$a = b \qquad a$$
$$S \quad \text{hence} \quad S$$
$$c \qquad\qquad c.$$

Inference is seen to take place in exactly the same manner as before by the substitution of a for b, and the negative equation or difference $b \backsim c$ is the part in which substitution takes place, but which has itself no substitutive power. Accordingly we shall in vain throw two differences into the same form, as in

$$a \backsim b \qquad b \backsim c$$
$$S \quad \text{or} \quad S$$
$$c \qquad\qquad a,$$

because we have no copula allowing us to make any substitution.

17. I am confirmed in this view by observing that, while the instrument of substitution is always an equation, the forms of relation in which a substitution may be made are by no means restricted to relations of equality or difference. If $a = b$, then in whatever way a third quantity c is related to one of them, in the same way it must be related to the other. If we take the sign ∞ to denote any conceivable kind of relation between one quantity and another, then the widest possible expression of a process of mathematical inference is shown in the form—

$$\begin{array}{ccc} a = b & & a \\ \S & \text{hence} & \S \\ c & & c. \end{array}$$

If in one case we take the sign ∞ as denoting that c is a multiple of b, it follows that it is a multiple of a; if it is the nth multiple of one, it is the nth multiple of the other; if it is the nth submultiple, or the nth power, or the nth root of one, it similarly follows that it stands in the same relation to the other; or if, lastly, c be greater than b by n or less than c by n, it will also be greater or less than a by n. In this all-powerful form we actually seem to have brought together the whole of the processes by which equations are solved, viz. equal addition or subtraction, multiplication or division, involution or evolution, performed upon both sides of the equation at the same time. That most familiar process in mathematical reasoning, of substituting one member of an equation for the other, appears to be the type of all reasoning, and we may fitly name this all-important process the *substitution of equals*.

18. An apparent exception to the statement that all mathematical reasoning proceeds by equations may perhaps occur to the reader, in the fact that reasoning can be conducted by *inequalities*. A chapter on the subject of inequalities may even be found in most elementary works on algebra, and it is self-evident that *a greater of a greater is a greater*, and *what is less than a less is less*. Thus we certainly seem to have in the two formulæ,

$$a > b > c \text{ hence } a > c,$$

and

$$a < b < c \text{ hence } a < c,$$

two valid modes of reasoning otherwise than by equations. But it is apparent, in the first place, that the use of these signs < and > demands some precautions which do not attach to the copula =; the formulæ,

$$a > b < c,$$
$$a < b > c,$$

do not establish any relation between a and c; and I think the reader will not find it easy to explain why these do not and the former do, without implying the use of an equation or identity. The truth is, that the formulæ,

$$a > b > c,$$
$$a < b < c,$$

involve not only two differences, but also one identity in the direction of those differences, whereas the formulæ,

$$a > b < c,$$
$$a < b > c,$$

appear to fail in giving any inference because they involve only differences both of direction and quantity.

Strength is added to this view of the matter by observing that all reasoning by inequalities can be represented with equal or superior clearness and precision in the form of *equalities*, while the contrary is by no means always true. Thus the inequality

$$a > b$$

is represented by the equality

(1) $a = b + p,$

THE TRUE PRINCIPLE OF REASONING

in which p is any positive quantity greater than zero; and the inequality

$$b > c$$

is similarly represented by the equality

(2) $\quad b = c + q,$

in which q is again a positive quantity greater than zero. By substituting for b in (1) its value as given in (2), we obtain the equation

$$a = c + p + q,$$

which, owing to the like signs of p and q, is a representation in a more exact and clear manner of the conclusion

$$a > c.$$

On the other hand, the formula

$$a > b < e$$

would evidently lead to the equation

$$a = e + p - r,$$

in which p is the excess of a over b, and r the excess of e over b. Now this equation, taken in connection with the former one, seems to give much clearer information as to the conditions under which inference is possible than do the formulæ of inequalities, and I entertain no doubt at all that, even when an inference seems to be obtained without the use of an equation, a disguised substitution is really performed by the mind, exactly such as represented in the equations. But I can only assert my belief of this from the examination of the process in my own mind, and I must submit to the reader's judgment whether there are exceptions or not to the rule, that we always reason by means of identities or equalities.

19. Turning now to apply these considerations to the forms of logical inference, my proposed simplification of the rules of logic is founded upon an obvious extension of the one great process of substitution to all kinds of identity. The Latin word *æqualis*, which is the original of our *equal*, was not restricted in signification to similarity of quantities, but was often applied to anything which was unvaried or similar when compared with another. We have but to interpret the word *equal* in the older and wider sense of like or equivalent, in order to effect the long-desired union of logical and mathematical reasoning. For it is not difficult to show that all forms of reasoning consist in repeated employment of the universal process of the *substitution of equals*, or, if the phrase be preferred, *substitution of similars*.

20. To prevent a confusion of mathematical and logical applications of the formula, it will be desirable to use large capital letters to denote the things compared in a logical sense, but the copula or sign of identity may remain as before. Thus the symbols.

$$A = B$$

denote the identity of the things represented by the indefinite terms or names A and B. Thus A may be taken in one case to mean *Iron*, when B might mean *the cheapest of the metals*, or *the most useful of the metals*. In another example which we have used A would mean *element*, and B *that which cannot be decomposed*, and so on. The fundamental principle of reasoning authorises us to substitute the term on one side of an identity for the other term, *wherever this may be encountered*, so that in whatever relation B stands to a third thing C, in the same relation A must stand to C. Or, using the sign ∞ to denote any possible or conceivable kind of relation, the formula

$$\begin{array}{cc} A = B & A \\ \S & \text{hence} \quad \S \\ C & C \end{array}$$

represents a self-evident inference. Thus,

If C *be the father of* B, C *is father of* A;
If C *be a compound of* B, C *is a compound of* A;
If C *be the absence of* B, C *is the absence of* A;
If C *be identical with* B, C *is identical with* A;

and so on.

21. We may at once proceed to develop from this process of substitution all the forms of inference recognised by Aristotle, and many more. In the first place, there cannot be a simpler act of reasoning than the substitution of a definition for a term defined; and though this operation found no place in the old system of the syllogism, it ought to hold the first place in a true system. If we take the definition of element as

$$Element = undecomposable\ substance,$$

we are authorised to employ the terms *element* and *undecomposable substance* in lieu of each other in whatever relation either of them may be found. If we describe iron as a *kind of element*, it may also be described as a *kind of undecomposable substance*.

22. Sometimes we may have two definitions of the same term, and we may then equate these to each other. Thus, according to Mr. Senior,

(1) *Wealth = whatever has exchangeable value.*

(2) *Wealth = whatever is useful, transferable, and limited in supply.*

We can employ either of these to make a substitution in the other, obtaining the equation,

Whatever has exchangeable value = whatever is useful, transferable, and limited in supply.

Where we have one affirmative proposition or equation, and one negative proposition, we still find the former sufficient for the process of inference. Thus—

(1) *Iron = the most useful metal.*

(2) *Iron* ᓕ *the metal most early used by primitive nations.*

By substituting in (2) by means of (1) we have

The most useful metal ∽ *the metal most early used by primitive nations.*

23. But two negative propositions will of course give no result. Thus the two propositions,

Snowdon ∽ *the highest mountain in Great Britain,*
Snowdon ∽ *the highest mountain in the world,*

do not allow of any substitution, and therefore do not give any means of inferring whether or not the highest mountain in Great Britain is the highest mountain in the world.

24. Postponing to a later part of this tract (§ 36) the consideration of negative forms of inference, I will now notice some inferences which involve combinations of terms. However many nouns, substantive or adjective, may be joined together, we may substitute for each its equivalent. Thus, if we have the propositions,

$$Square = equilateral\ rectangle,$$
$$Equilateral = equal\text{-}sided,$$
$$Rectangle = right\text{-}angled\ quadrilateral,$$
$$Quadrilateral = four\text{-}sided\ figure,$$

we may by evident substitutions obtain

Square = *equal-sided, right-angled, four-sided figure.*

25. It is desirable at this point to draw attention to the fact that the order in which nouns adjective are stated is a matter of indifference. A *four-sided, equal-sided figure* is identically the same as *an equal-sided, four-sided figure;* and even when it sometimes seems inelegant or difficult to alter the order of names describing a thing, it is grammatical usage, not logical necessity, which stands in the way. Hence, if A and B represent any two names or terms, their junction as in AB will be taken to indicate anything which unites the qualities of both A and B, and then it follows that

$$AB = BA.$$

This principle of logical symbols has been fully explained by Dr. Boole in his *Laws of Thought* (pp. 29, 30), and also in my *Pure Logic* (p. 15); and its truth will be assumed here without further proof. It must be observed, however, that this property of logical symbols is true only of adjectives, or their equivalents, united to nouns, and not of words connected together by prepositions, or in other ways. Thus *table of wood* is not equivalent to *wood of table*; but if we treat the words *of wood* as equivalent to the adjective *wooden*, it is true that a *table of wood* is the same as a *wooden table*.

26. We may now proceed to consider the ordinary proposition of the form

$$A = AB,$$

which asserts the identity of the class A with a particular part of the class B, namely the part which has the properties of A. It may seem when stated in this way to be a truism, but it is not, because it really states in the form of an identity the inclusion of A in a wider class B. Aristotle happened to treat it in the latter aspect only, and the extreme incompleteness of his syllogistic system is due to this circumstance. It is only by treating the proposition as an identity that its relation to the other forms of reasoning becomes apparent.

27. One of the simplest and by far the most common form of argument in which the proposition of the above form occurs is the mood of the syllogism known by the name *Barbara*.

As an example, we may take the following:—

(1) *Iron is a metal*,
(2) *A metal is an element*, therefore
(3) *Iron is an element*.

The propositions thus expressed in the ordinary manner become, in a strictly logical form—

(1) $Iron = metallic\ iron$,
(2) $Metal = elementary\ metal.$

Now for *metal* or *metallic* in (1) we may substitute its equivalent in (2) and we obtain

(3) *Iron = elementary, metal, iron ;*

which in the elliptical expression of ordinary conversation becomes *Iron is an element,* or Iron is *some kind* of element, the words *an* or *some kind* being indefinite substitutes for a more exact description

The form of this mode of inference must be stated in symbols on account of its great importance. If we take

$$A = iron,$$
$$B = metal,$$
$$C = element,$$

the premises are obviously,

(1) $A = AB$,
(2) $B = BC$,

and substituting for B in (1) its description in (2) we have the conclusion

$$A = ABC,$$

which is the symbolic expression of (3).

28. The mood *Darii*, which is distinguished from *Barbara* in the doctrine of the syllogism by its particular minor premise and conclusion, cannot be considered an essentially different form. For if, instead of taking A in the previous example = *iron*, we had taken it

A = *some native minerals ;*

B and C remaining as before, we should then have the conclusion

$$A = ABC,$$

denoting

some native minerals are elements ;

which affords an instance of the syllogism Darii exhibited in exactly the same form as Barbara.

THE TRUE PRINCIPLE OF REASONING

29. The *sorites* or chain of syllogisms consists but in a series of premises of the same kind, allowing of repeated substitution. Let the premises be—

      (1) *The honest man is truly wise,*
      (2) *The truly wise man is happy,*
      (3) *The happy man is contented,*
      (4) *The contented man is to be envied,*

the conclusion being—

      (5) *The honest man is to be envied.*

Taking the letters A, B, C, D, and E to indicate respectively *honest* man, *truly wise, happy, contented,* and *to be envied,* the premises are represented thus—

      (1) $A = AB$,
      (2) $B = BC$,
      (3) $C = CD$,
      (4) $D = DE$,

and successive substitutions by (4) in (3), by (3) in (2), and by (2) in (1), give us

$$C = CDE,$$
$$B = BCDE,$$
$$A = ABCDE.$$

Or we may get exactly the same conclusion by substitution in a different order, thus—

$$A = AB = ABC = ABCD = ABCDE.$$

The ordinary statement of the conclusion in (5) is only an indefinite expression of the full description of A given in $A = ABCDE$.

30. All the affirmative moods of the syllogism may be represented with almost equal clearness and facility. As an example of *Darapti* in the third figure we may take

      (1) *Oxygen is an element,*
      (2) *Oxygen is a gas,*
      (3) *Some gas, therefore, is an element.*

Making $\quad$ A = *gas*,
$\qquad\qquad\qquad$ B = *oxygen*,
$\qquad\qquad\qquad$ C = *element*,
the premises become

$\qquad$ (1) B = BC,
$\qquad$ (2) B = BA.

Hence, by obvious substitution, either by (1) in (2) or by (2) in (1), we get

$\qquad$ (3) BA = BC.

Precisely interpreted this means that *gas which is oxygen* is *element which is oxygen;* but when this full interpretation is unnecessary, we may substitute the indefinite adjective *some* for the more particular description, getting,

$\qquad$ *Some gas is some element,*

or, in the still more vague form of common language,

$\qquad$ *Some gas is an element.*

31. The mood *Datisi* may thus be illustrated—

$\qquad$ (1) *Some metals are inflammable,*
$\qquad$ (2) *All metals are elements,*
$\qquad$ (3) *Some elements are inflammable.*

Taking
$\qquad$ A = *elements*, $\qquad$ C = *inflammable*,
$\qquad$ B = *metals*, $\qquad\;\,$ D = *some*,

we may represent the premises in the forms

$\qquad$ (1) DB = DBC
$\qquad$ (2) $\;\,$B = BA.

Substitution, in the second side of (1), of the description of B given in (2) produces the conclusion

$\qquad$ (3) DB = DBCA,

or, in words,

$\qquad$ *Some metal = some metal element inflammable.*

In this and many other instances my method of representation is found to give a far more full and strict conclusion than the old syllogism; but ellipsis or a substitution of indefinite particles or adjectives easily enables us to pass from the strict form to the vague results of the syllogism: it would be in vain that we should attempt to reach the more strict conclusion by the syllogism alone. But I must beg of the reader not to judge the validity of my forms by any single instance only, but rather by the wide embracing powers of the principle involved. Even common thought must be condemned as loose and imperfect if it should be found in certain cases to be inconsistent with a generalisation which holds true throughout the exact sciences as well as the greater part of the ordinary acts of reasoning.

32. Certain forms of so-called *immediate inference*, chiefly brought into notice in recent times by Dr. Thomson, are readily derived from our principle.

Immediate inference by added determinant[1] consists in joining a determining or qualifying adjective, or some equivalent phrase, to each member of a proposition, a new proposition being thus inferred. Dr. Thomson's own example is as follows—

A negro is a fellow-creature;

whence we infer immediately,

A negro in suffering is a fellow-creature in suffering.

To explain accurately the mode in which this inference seems to be made according to our principle, let us take

$A = negro,$
$B = fellow\text{-}creature,$
$C = suffering.$

The premise may be represented as

$$A = AB.$$

Now it is self-evident that AC is identical with AC, this

[1] *Outline of the Laws of Thought,* § 87.

being a fact which some may think to be somewhat unnecessarily laid down in the first of the primary laws of thought (see § 41).

In the symbolic expression of this fact,

$$AC = AC,$$

we can substitute for A in the second member its equivalent AB, getting

$$AC = ABC.$$

This may be interpreted in ordinary words as,

A suffering negro is a suffering negro fellow-creature,

which differs only from the conclusion as stated by Dr. Thomson by containing the qualification *negro* in the second member.

33. *Immediate inference by complex conception* closely resembles the preceding, and is of exceedingly frequent occurrence in common thought and language, although it has never had a properly recognised place in logical doctrine until lately.[1]

Its nature is best learnt from such an example as the following:—

Oxygen is an element,
Therefore a pound weight of oxygen is a
pound weight of an element.

This is a very plain case of substitution; for if we make

$$O = oxygen,$$
$$P = pound\ weight,$$
$$Q = element,$$

we may represent the premise as

$$O = OQ.$$

Now it is self-evident that

$$P\ of\ O = P\ of\ O,$$

[1] Thomson's *Outline*, § 88.

and substituting in the second member the description of O we have

$$P \ of \ O = P \ of \ OQ.$$

34. In an exactly similar manner we may solve a common form of reasoning which the authors of the *Port Royal Logic* described as the *Complex Syllogism*, remarking how little attention logicians had in their day given to many common forms of reasoning.[1] I will employ their example, which is as follows—

(1) *The sun is an insensible thing,*
(2) *The Persians worship the sun,*
(3) *The Persians, therefore, worship an insensible thing.*

Making

$$A = sun,$$
$$B = insensible \ thing,$$
$$C = Persians,$$
$$D = worshippers,$$

we may represent the above by the symbols

(1) $A = AB$
(2) $C = CD \ of \ A.$

Hence, by substitution for A in (2) by means of (1),

(3) $C = CD \ of \ AB.$

35. I regard hypothetical propositions as only differing from categorical propositions in the accidental form of expression. It is well known to readers of the ordinary handbooks of logic, that hypothetical propositions can always be represented in the categorical form by altering the phraseology; and the fact that the alteration required is often of the slightest possible character seems to show that there is no essential difference. Thus the proposition,

If iron contain phosphorus, it is brittle,

[1] *Port Royal Logic*, translated by Mr. Spencer Baynes, p. 207.

is hypothetical, but exactly equivalent to the categorical proposition,

Iron, containing phosphorus, is brittle;

which is of the symbolic form,

$$AB = ABC.$$

But propositions such as,

If the barometer falls, a storm is coming,

cannot be reduced but by some such mode of expression as the following:—

The circumstances of a falling barometer are the circumstances of a storm coming.

Nevertheless, sufficient freedom in the alteration of expression being granted, they readily come under our formulæ.

36. I have as yet introduced few examples of negative propositions, because, though they may be treated in their purely negative form, it is usually more convenient to convert them into affirmative propositions. This conversion is effected by the use of *negative terms*, a practice not unknown to the old logic, but not nearly so much employed as it should have been. Thus the negative proposition,

$$A \text{ is not } B$$
or
$$A \backsim B,$$

is much more conveniently represented by the affirmative proposition or equation,

$$A = Ab,$$

in which we denote by b the quality or fact of differing from B. The term b is in fact the name of the whole class of things, or any of them, which differ from B, so that it is a matter of indifference whether we say that A differs from B and is excluded from the class B, or that it agrees with b and is included in the class b. There are advantages, however, in employing the affirmative form.[1]

37. The syllogism *Celarent* is now very readily brought under our single mode of inference. Take the example

(1) *All metals are elements,*
(2) *No element can be transmuted,*
(3) *No metal, therefore, can be transmuted.*

To represent this symbolically, let

$A = metal,$
$B = element,$
$C = transmutable,$
$c = untransmutable.$

Then the premises are

(1) $A = AB$
(2) $B = Bc.$

Substituting in (1) by means of (2) we get

(3) $A = ABc;$

or, *metals = metals, elementary, untransmutable.*

38. Before proceeding to other examples of the syllogism,

---

[1] It may seem to the reader contradictory to condemn the negative proposition as sterile and incapable of affording inferences, and shortly afterwards to convert it into an affirmative proposition of fertile or inferential power. But on trial it will be found that the propositions thus obtained yield no conclusions inconsistent with my theory. Thus the negative premises,

A *is not* $B,$
B *is not* $C,$

yield the affirmative propositions or equations,

$A = Ab$
$B = Bc.$

And when these premises are tested, whether on the logical slate, abacus, or logical machine referred to in a later page, they are found to give no conclusion concerning the relation of A and C. The description of A is given in the equation,

$$A = AbC \cdot |\cdot Abc,$$

from which it appears that A may indifferently occur with or without C.

it will be well to point out that every affirmative proposition or equation gives rise to a corresponding equation between the negatives of the terms of the original. The general proposition of the form

$$A = B,$$

treated by the fundamental principle of reasoning, informs us that in whatever relation anything stands to A, in the same relation it stands to B, and similarly *vice versâ*. Hence, whatever differs from A differs from B, and whatever differs from B differs from A. Now the term b denotes what differs from B, and a denotes what differs from A; so that from the single original proposition we may draw the two propositions—

$$a = ab$$
$$b = ab.$$

But as these propositions have an identical second member, we can make a substitution, getting

$$a = b.$$

This form of inference, though little if at all noticed in the traditional logic, is of frequent occurrence and of great importance. It may be illustrated by such examples as

happy = contented,

hence *unhappy = not-contented ;*

or again, *triangle = three-sided rectilinear figure,*

hence

what is not a triangle = what is not a three-sided rectilinear figure.

The new proposition thus obtained may be called the *contrapositive* of the one from which it was derived, this being a name long applied to a similar inference from the old form of proposition.

39. Though the details of this new view of logic may not yet have been perfectly worked out, much evidence of the truth of the system is to be found in the simplicity, variety, and universality of the forms of reasoning which can be evolved out of a single law of thought,—the *similar treatment of similars*.

The old system of the syllogism, indeed, was nominally founded on a single, or rather double, axiom or law, the *dicta* of Aristotle, but the mode in which these *dicta* led to conclusions was so far from being evident, that the logical student could not be trusted with their use. A cumbrous system of six, eight, or more rules of the syllogism was therefore made out, in order that the validity of an argument might thereby be tested; but, as even then the task was no easy or self-evident one, logicians formed a complete list of the limited number of forms obeying these artificial rules, and composed a curious set of mnemonic lines by which they might be committed to memory. These lines, the venerable *Barbara, Celarent*, etc., were no doubt creditable to the ingenuity of men who lived in the darkest ages of science, but they are altogether an anachronism in the present age. What should we think now of a writer of mathematical textbooks, who should select about a score of the commonest forms of mathematical equations, and invent a mnemonic by which both the forms of the equations and the steps of their solution might be carried in the memory? Instead of such an absurdity, we now find, even in purely elementary books, that the general principles and processes are impressed upon the pupil's mind, and he is taught by practice to apply these principles to indefinitely numerous and varied examples. So it should be in logic; the logical student need only acquire a thorough comprehension of the principle of substitution and the very primary laws of thought, in order to be able to analyse any argument and develop any form of reasoning which is possible. No subsidiary rules are needful, and no mnemonics would be otherwise than a hindrance.

40. I have yet a striking proof to offer of the truth of the views I am putting forward; for when once we lay down the primary laws of thought, and employ them by means of the principle of substitution, we find that an unlimited system of forms of indirect reasoning develops itself spontaneously. Of this indirect system there is hardly a vestige in the old logic, nor does any writer previous to Dr. Boole appear to have conceived its existence, though it must no doubt have been often unconsciously employed in particular cases. This indirect or negative method is closely analogous to the *indirect proof,* or *reductio ad absurdum,* so frequently used by Euclid and other mathematicians, and a similar method is employed by the old logicians in the treatment of the syllogisms called *Baroko* and *Bokardo,* by the *reductio ad impossibile.* But the incidental examples of the indirect logical method which can be found in any book previous to the *Mathematical Analysis of Logic* of Dr. Boole give no idea whatever of its all-commanding power; for it is not only capable of proving all the results obtained already by a direct method of inference, but it gives an unlimited number of other inferences which could not be arrived at in any other than a negative or indirect manner. In a previous little work[1] I have given a complete, but somewhat tedious, demonstration of the nature and results of this method, freed from the difficulties and occasional errors in which Dr. Boole left it involved. I will now give a brief outline of its principles.

41. The indirect method is founded upon the law of the substitution of similars as applied with the aid of the fundamental laws of thought. These laws are not to be found in most textbooks of logic, but yet they are necessarily the basis of all reasoning, since they enounce the very nature of similarity or identity. Their existence is assumed or implied, therefore, in the complicated rules of the syllogism, whereas my system is founded upon an immediate application of the laws themselves. The first of these laws, which I have already referred to in an earlier part of this tract (p. 103), is

---

[1] *Pure Logic, or the Logic of Quality apart from Quantity: with Remarks on Boole's System, and on the Relation of Logic and Mathematics.* By W. Stanley Jevons, M.A. London: Edward Stanford, 1864.

the LAW OF IDENTITY, that *whatever is, is*, or *a thing is identical with itself;* or, in symbols,

$$A = A.$$

The second law, THE LAW OF NON-CONTRADICTION, is that *a thing cannot both be and not be*, or that *nothing can combine contradictory attributes;* or, in symbols,

$$Aa = 0,$$

—that is to say, what is both A and not A does not exist, and cannot be conceived.

The third law, that of *excluded middle*, or, as I prefer to call it, the LAW OF DUALITY, asserts the self-evident truth that *a thing either exists or does not exist*, or that *everything either possesses a given attribute or does not possess it.*

Symbolically the law of duality is shown by

$$A = AB \dotplus Ab,$$

in which the sign $\dotplus$ indicates alternation, and is equivalent to the true meaning of the disjunctive conjunction *or*. Hence the symbols may be interpreted as, A *is either* B *or not* B.

These laws may seem truisms, and they were ridiculed as such by Locke; but, since they describe the very nature of identity in its three aspects, they must be assumed as true, consciously or unconsciously, and if we can build a system of inference upon them, their self-evidence is surely in our favour.

42. The nature of the system will be best learnt from examples, and I will first apply it to several moods of the old syllogism. *Camestres* may thus be proved and illustrated—

(1) *A sun is self-luminous,*
(2) *A planet is not self-luminous,*
(3) *A planet, therefore, is not a sun.*

Now it is apparent that a planet is either a sun or it is not a sun, by the law of duality. But if it be a sun, it is self-luminous by (1), whereas by (2) it is not self-luminous; it would, if a sun, combine contradictory attributes. By the law of non-contradiction it could not exist, therefore, as a sun, and it consequently is not a sun.

To represent this reasoning in symbols take

$$A = sun.$$
$$B = planet.$$
$$C = self\text{-}luminous.$$

Then the premises are

(1) $A = AC$
(2) $B = Bc.$

By the law of duality we have

$$B = BA \cdot|\cdot Ba,$$

and substituting this value in the second side of (2) we have

$$B = BcA \cdot|\cdot Bca.$$

But for A in the above we may substitute its expression in (1), getting

$$B = BcAC \cdot|\cdot Bca ;$$

and striking out one of these alternatives which is contradictory we finally obtain

$$B = Bca.$$

The meaning of this formula is that *a planet is a planet not self-luminous, and not a sun,* which only differs from the Aristotelian conclusion in being more full and precise.

43. The syllogism *Camenes* may be illustrated by the following example:—

(1) *All monarchs are human beings,*
(2) *No human beings are infallible,*
(3) *No infallible beings, therefore, are monarchs.*

This is proved by considering that every infallible being is either a monarch or not a monarch; but if a monarch, then by (1) he is a human being, and by (2) is not infallible, which is impossible; therefore, no infallible being is a monarch.

Or in symbols, taking

$$A = monarch,$$
$$B = human\ being,$$
$$C = infallible\ being,$$

the premises are

(1) $A = AB$
(2) $B = Bc.$

Now by the law of duality

$$C = aC \cdot\!\!|\cdot AC.$$

Substituting for A its value as derived from both the premises, we have

$$C = aC \cdot\!\!|\cdot ABCc\ ;$$

and, striking out the contradictory term,

$$C = aC.$$

44. By the indirect method we can obtain and prove the truth of the contra-positive of the ordinary proposition A is B, or

(1) $A = AB.$

What we require is the description of the term not-B or b; and by the law of duality this is, in the first place, either A or not-A:

(2) $b = Ab \cdot\!\!|\cdot ab.$

Substituting for A in (2) its value as given in (1) we obtain

$$b = ABb \cdot\!\!|\cdot ab.$$

But the term ABb breaks the law of non-contradiction (p. 111), so that we have left only

$$b = ab,$$

or whatever is not-B is also not-A.

Thus, if   A = *metal*,
    B = *element*,

from the premise

All metals are elements

we conclude *that all substances which are not-elements are not metals;* which is proved at once by the consideration, symbolically expressed above, that if they were metals they would be elements, or *at once elements and not-elements*, which is impossible.

45. It is the peculiar character of this method of indirect inference that it is capable of solving and explaining, in the most complete manner, arguments of any degree of complexity. It furnishes, in fact, a complete solution of the problem first propounded and obscurely solved by Dr. Boole—

Given any number of propositions involving any number of distinct terms, required the description of any of those terms or any combination of those terms as expressed in the other terms, under condition of the premises remaining true.

This method always commences by developing all the possible combinations of the terms involved according to the law of duality. Thus, if there are three terms, represented by A, B, C, then the possible combinations in which A can present itself will not exceed four, as follows—

(1) A = ABC ·|· AB*c* ·|· A*b*C ·|· A*bc*.

If we have any premises or statements concerning the nature of A, B, and C, that is, the combinations in which they can present themselves, we proceed to inquire how many of the above combinations are consistent with the premises. Thus, if A is never found with B, but B is always found with C, the two first of the combinations become contradictory, and we have

$$A = AbC \cdot | \cdot Abc,$$

or, A is never found with B, but may or may not be found with C.

This conclusion may be proved symbolically by expressing the premises thus—

$$A = Ab$$
$$B = BC,$$

and then substituting the values of A and B wherever they occur on the second side of (1).

46. As a simple example of the process, let us take the following premises, and investigate the consequences which flow from them.[1]

'From A follows B, and from C follows D; but B and D are inconsistent with each other.'

The possible combinations in which A, B, C, and D may present themselves are sixteen in number, as follows—

A B C D	a B C D
A B C d	a B C d
A B c D	a B c D
A B c d	a B c d
A b C D	a b C D
A b C d	a b C d
A b c D	a b c D
A b c d	a b c d

Each of these combinations is to be compared with the premises in order to ascertain whether it is possible under the condition of those premises. This comparison will really consist in substituting for each letter its description as given in the premises, which may thus be symbolically expressed—

(1) $A = AB$,
(2) $C = CD$,
(3) $B = Bd$.

The combination A b C D is contradicted by (1) in substituting for A its value; A B C d by (2), a B c D by (3), and

[1] See De Morgan's *Formal Logic*, p. 123.

so on. There will be found to remain only four possible combinations—

$$A\,B\,c\,d,$$
$$a\,b\,C\,D,$$
$$a\,b\,c\,D,$$
$$a\,b\,c\,d.$$

Now, if we wish to ascertain the nature of the term A, we learn at once that it can only exist in the presence of B and the absence of both C and D.

We ascertain also that D can only appear in the absence of both A and B, but that C may or may not be present with D. Where D is absent, C must also be absent, and so on.

47. Objections might be raised against this process of indirect inference, that it is a long and tedious one; and so it is, when thus performed. Tedium indeed is no argument against truth; and if, as I confidently assert, this method gives us the means of solving an infinite number of problems, and arriving at an infinite number of conclusions, which are often demonstrable in no simpler way, and in fact in no other way whatever, no such objections would be of any weight. The fact however is, that almost all the tediousness and liability to mistake may be removed from the process by the use of mechanical aids, which are of several kinds and degrees. While practising myself in the use of the process, I was at once led to the use of the *logical slate*, which consists of a common writing slate, with several series of the combinations of letters engraved upon it, thus—

A B	A B C	A B C D	A B C D E	A B C D E F
A b	A B c	A B C d	A B C D e	A B C D E f
a B	A b C	A B c D	A B C D e	A B C D e F
a b	A b c	A B c d	A B C d e	A B C D e f
	a B C	A b C D	A B c D E	A B C d E F
	a B c	...	...	...
	a b C	...	...	...
	a b c	a b c d	a b c d e	a b c d e f

THE TRUE PRINCIPLE OF REASONING

When fully written out, these series consist respectively of 4, 8, 16, 32, and 64 combinations, and that series is chosen for any problem which just affords enough distinct terms. Each combination is then examined in connection with each of the premises, and the contradictory ones are struck through with the pencil.

48. It soon became apparent, however, that if these combinations, instead of being written in fixed order on a slate, were printed upon light movable slips of wood, it would become easy by suitable mechanical arrangements to pick out the combinations in convenient classes, so as immensely to abbreviate the labour of comparison with the premises. This idea was carried out in the *logical abacus*, which I constructed several years ago, and have found useful and successful in the lecture-room for exhibiting the complete solution of logical arguments.

49. This *logical abacus* has been exhibited before the members of the Manchester Literary and Philosophical Society, and the following description of it is extracted from the *Proceedings* of the Society for 3d April, 1866, p. 161.

'The abacus consists of—

'1. An inclined black board, furnished with four ledges, 3 ft. long, placed 9 in. apart.

'2. Series of flat slips of wood, the smallest set four in number, and other sets, 8, 16, and 32 in number, marked with combinations of letters, as follows—

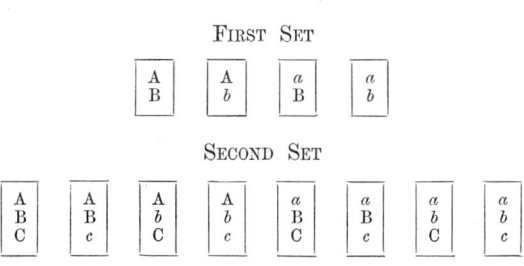

'The third and fourth sets exhibit the corresponding

combinations of the letters A, B, C, D, *a, b, c, d*, and A, B, C, D, E, *a, b, c, d, e*.

'The slips are furnished with little pins, so that, when placed upon the ledges of the board, those marked by any given letter may be readily picked out by means of a straight-edged ruler, and removed to another ledge.

50. 'The use of the abacus will be best shown by an example. Take the syllogism in *Barbara*—

> *Man is mortal,*
> *Socrates is man,*
> *Therefore Socrates is mortal.*

'Let
$$A = Socrates,$$
$$B = man,$$
$$C = mortal.$$

'The corresponding small italic letters then indicate the negatives,
$$a = not\text{-}Socrates,$$
$$b = not\text{-}man,$$
$$c = not\text{-}mortal,$$

and the premises may be stated as

$$A \text{ is } B,$$
$$B \text{ is } C.$$

'Now take the second set of slips containing all the possible combinations of A, B, C, *a, b, c*, and ascertain which of the combinations are possible under the conditions of the premises.

'Select all the slips marked A; and as all these ought to be B's, select again those which are not-B or *b*, and reject them. Unite the remainder, and, selecting the B's, reject those which are not-C or *c*. There will now remain only four slips or combinations—

A	*a*	*a*	*a*
B	B	*b*	*b*
C	C	C	*c*

'If we require the description of *Socrates*, or A, we take the only combination containing A, and observe that it is joined with C: hence the Aristotelian conclusion, *Socrates is mortal*. We may also get any other possible conclusion. For instance, the class of things *not-man* or b is seen from the two last combinations to be always *a* or *not-Socrates*, but either *mortal* or *not-mortal* as the case may be.

51. 'Precisely the same obvious system of analysis is applicable to arguments however complicated. As an example, take the premises treated in Boole's *Laws of Thought*, p. 125.

'"(1.) *Similar figures consist of all whose corresponding angles are equal, and whose corresponding sides are proportional.*"

'"(2.) *Triangles whose corresponding angles are equal have their corresponding sides proportional, and* vice versâ."

'Let

A = *similar*,
B = *triangle*,
C = *having corresponding angles equal*,
D = *having corresponding sides proportional*.

'The premises may then be expressed in Qualitative Logic, as follows—

$$A = CD,$$
$$BC = BD.$$

'Take the set of 16 slips: out of the A's reject those which are not CD; out of the CD's reject those which are not A; out of the BC's reject those which are not BD; and out of the BD's reject those which are not BC. There will remain only six slips, as follows—

A	A	a	a	a	a
B	b	B	b	b	b
C	C	c	C	c	c
D	D	d	d	D	d

'From these we may at once read off all the conclusions

laboriously deduced by Boole in his obscure processes. We at once see, for instance, that the class a, or "*dissimilar figures*, consist of *all triangles* (B) *which have not their corresponding angles equal* (*c*) *and sides proportional* (*d*), *and of all figures not being triangles* (*b*) *which have either their angles equal* (C) *and sides not proportional* (*d*), *or their corresponding sides proportional* (D) *and angles not equal, or neither their corresponding angles equal nor corresponding sides proportional.*" (Boole, p. 126).

52. 'The selections as made upon the abacus are of course subject to mistake, but only one easy step is required to a logical machine, in which the selections shall be made mechanically and faultlessly by the mere reading down of the premises upon a set of keys, or handles, representing the several positive and negative terms, the copula, conjunctions, and stops of a proposition.'

53. In the last paragraph I alluded to a further mechanical contrivance, in which the combination-slips of the abacus should not require to be moved by hand, but could be placed in proper order by the successive pressure of a series of keys or handles. I have since made a successful working model of this contrivance, which may be considered *a machine capable of reasoning,* or of replacing almost entirely the action of the mind in drawing inferences. When I have an opportunity of describing the details of its construction, I think it will be found to afford a physical proof, apparent to the eyes, of the extreme incompleteness of the Aristotelian logic. Not only are the syllogisms and other old forms of argument capable of being worked upon the machine, but an indefinite number of other forms of reasoning can be represented by the simple regular action of levers and spindles.

54. The most unfortunate feature of the long history of our present traditional logic has been the divorce existing between the logic of the schools and the logic of common life. There has been no apparent connection whatever between the formal strictness of the syllogistic art and the more loose but useful suggestions of analogy from particu-

THE TRUE PRINCIPLE OF REASONING 121

lars to particulars. It is owing to this separation, as I apprehend, that a succession of English writers from Locke down to Mr. J. S. Mill have been led to under-estimate the value of the syllogism. In Mr. Mill's system of logic the syllogism occupies a very anomalous position—that of an extraneous form of proof which may be employed when we wish to ensure correctness of inference, but which is useless for the discovery of truth. I believe that the new view of the syllogism which I am now proposing will remedy this lamentable disconnection of the parts of what should be one most harmonious and consistent whole. There is no subject in which we might expect more perfect unity and system to exist, and more wide-ruling generalisations to be discoverable, than in the science of the laws of thought; and I conceive that a prime object of any logical reform should be to reconcile the strict doctrine with the looser forms of ordinary thought. This reconciliation will really be effected, I believe, by adopting as the fundamental principle the modified axiom of Aristotle which I have called the *substitution of similars*. I hope at some future time to explain fully the results which seem to follow from the principle and the harmony which it creates between the several branches of logical method, and I will only attempt in this tract a few slight illustrations.

55. The most frequent mode of inference in common life is that known as reasoning from analogy or resemblance, by which we argue from any thing or event we have known to a like thing or event encountered on another occasion. This seems to be Mr. Mill's view of the ordinary process of reasoning, for in discussing the functions and value of the syllogism, he says:[1] 'From instances which we have observed, we feel warranted in concluding that what we found true in those instances holds in all similar ones, past, present, and future, however numerous they may be.' And again he explains more fully:[2] 'I believe that, in point of fact,

[1] *System of Logic*, vol. i, p. 210, fifth edition.
[2] *Ibid.* p. 212.

when drawing inferences from our personal experience, and not from maxims handed down to us by books or tradition, we much oftener conclude from particulars to particulars directly, than through the intermediate agency of any general proposition. We are constantly reasoning from ourselves to other people, or from one person to another, without giving ourselves the trouble to erect our observations into general maxims of human or external nature. When we conclude that some person will, on some given occasion, feel or act so and so, we sometimes judge from an enlarged consideration of the manner in which human beings in general, or persons of some particular character, are accustomed to feel and act; but much oftener from merely recollecting the feelings and conduct of the same person in some previous instance, or from considering how we should feel or act ourselves. It is not only the village matron who, when called to a consultation upon the case of a neighbour's child, pronounces on the evil and its remedy simply on the recollection and authority of what she accounts the similar case of her Lucy.'

56. Mr. Mill expresses as clearly as it is well possible that we argue in common life, as he thinks, not by the syllogism, but directly from instance to instance by the similarity observed between the instances. But this argument from similars to similars is the identical process which I have called the substitution of similars, and which I have shown to be capable of explaining the syllogism itself, and much more. In fact, we find Mr. Mill enunciating this principle himself in another chapter, where he is treating of argument from analogy or resemblance. After noticing the stricter meaning of analogy as a *resemblance of relations*, he continues [1]—

'It is on the whole more usual, however, to extend the name of analogical evidence to arguments from any sort of resemblance, provided they do not amount to a complete induction: without peculiarly distinguishing resemblance of

[1] *System of Logic*, vol. ii, p. 86.

relations. Analogical reasoning, in this sense, may be reduced to the following formula: Two things resemble each other in one or more respects; a certain proposition is true of the one; therefore it is true of the other. But we have nothing here by which to discriminate analogy from induction, since this type will serve for all reasoning from experience. In the most rigid induction, equally with the faintest analogy, we conclude, because A resembles B in one or more properties, that it does so in a certain other property.'

57. If this be, as Mr. Mill so clearly states, the type of all reasoning from experience, it follows that the principle of inductive reasoning is actually identical with that which I have shown to be sufficient to explain the forms of deductive reasoning. The only difference I apprehend is, that in deductive reasoning we know or assume a similarity or identity to be certainly known, and the conclusion from it is therefore equally certain; but in inductive arguments from one instance to another we never can be sure that the similarity of the instance is so deep and perfect as to warrant our substitution of one for the other. Hence the conclusion is never certain, and possesses only a degree of probability, greater or less according to the circumstances of the case; and the theory of probabilities is our only resource for ascertaining this degree of probability, if ascertainable at all.

58. It is instructive to contrast mathematical induction with the induction as employed in the experimental sciences. The process by which we arrive at a general proof of a problem in Euclid's *Elements of Geometry* is really a process of generalisation presenting a striking illustration of our principle. To prove that the square on the hypothenuse of a right-angled triangle is equal to the sum of the squares on the sides containing the right angle, Euclid takes only a single example of such a triangle, and proves this to be true. He then trusts to the reader perceiving of his own accord that all other right-angled triangles resemble the one accidentally adopted in the points material to the proof, so that any one right-angled triangle may be indifferently substituted for any other. Here the process from one case to another is certain, because we *know* that one case exactly resembles another. In physical science it is not so, and the distinction has been expressed, as it seems to me, with admirable insight by Professor Bowen in his well-known *Treatise on Logic, or the Laws of Pure Thought*.[1] He says of mathematical figures: 'The same measure of certainty which the student of nature obtains by intuition respecting a single real object, the mathematician acquires respecting a whole class of imaginary objects, because the latter has the assurance, which the former can never attain, that the single object which he is contemplating in thought is a *perfect representative* of its whole class: he has this assurance, *because the whole class exists only in thought*, and are therefore all actually before him, or present to consciousness. For example: this bit of iron, I find by direct observation, melts at a certain temperature; but it may well happen that another piece of iron, quite similar to it in external appearance, may be fusible only at a much higher temperature, owing to the unsuspected presence with it of a little more or a little less carbon in composition. But if the angles at the base of this triangle are equal to each other, I know that a corresponding equality must exist in the case of every other figure which conforms to the definition of an isosceles triangle; for that definition excludes every disturbing element. The conclusion in this latter case, then, is universal, while in the former it can be only singular or particular.'

This passage perfectly supports my view that all reasoning consists in taking one thing as a *representative*, that is to say, as a *substitute*, for another, and the only difficulty is to estimate rightly the degree of certainty, or of mere probability, with which we make the substitution. The forms and methods of induction and the calculus of probabilities are necessary to guide us rightly in this; but to

[1] Cambridge, United States, 1866, p. 354.

show that the principle of substitution is really present and active throughout inductive logic is more than I can undertake to show in this tract, although I believe it to be so.

59. Though I have pointed out how consistent are many of Mr. Mill's expressions with the view of logic here put forward, and how clearly in one place he describes the principle of substitution itself, I cannot but feel that his system is full of anomalies and breaches of consistency. These arise, I believe, from the profound error into which he has fallen, of undervaluing the logical discovery of the quantification of the predicate. Of Sir W. Hamilton's views he says:[1] 'If I do not consider the doctrine of the quantification of the predicate a valuable accession to the art of logic, it is only because I consider the ordinary rules of the syllogism to be an adequate test, and perfectly sufficient to exclude all inferences which do not follow from the premises. Considered, however, as a contribution to the *science* of logic, that is, to the analysis of the mental processes concerned in reasoning, the new doctrine appears to me, I confess, not merely superfluous, but erroneous; since the form in which it clothes the propositions does not, like the ordinary form, express what is in the mind of the speaker when he enunciates the proposition. I cannot think Sir William Hamilton right in maintaining that the quantity of the predicate is "always understood in thought." It is implied, but is not present in the mind of the person who asserts the proposition.' Again, he says of Mr. De Morgan's ingenious logical discoveries, to which every logical writer is so deeply indebted: 'Since it is undeniable that inferences, in the cases examined by Mr. De Morgan, can legitimately be drawn, and that the ordinary theory takes no account of them, I will not say that it was not worth while to show in detail how these also could be reduced to formulæ as rigorous as those of Aristotle. What Mr. De Morgan has done was worth doing once (perhaps more than once), as a school exercise; but I question if its results are

[1] *System of Logic*, fifth ed. vol. i, p. 196, note.

worth studying and mastering for any practical purpose.' In these and many other places Mr. Mill shows a lamentable want of power of appreciating the principles involved in the quantification of the predicate. As regards the most original discoveries of Dr. Boole, there is not, so far as I have been able to discover, a single word in Mr. Mill's edition of his *Logic* published in 1862, to indicate that he was conscious of the publication of Mr. Boole's *Mathematical Analysis* in 1847, and of his great work, *The Laws of Thought*, in 1854. Although accounted a disciple and potent supporter of the doctrines of Jeremy Bentham, he appears unaware that the doctrine of the quantification of the predicate is traceable to his great master, or at all events to the work of a nephew founded upon the manuscripts of Bentham.

60. I ought not to omit to notice that Dr. Thomson substantially adopts the principle of substitution in treating of what he calls the *syllogism of analogy*. He states the canon in the following manner:[1] 'The same attributes may be assigned to distinct but similar things, provided they can be shown to accompany the points of resemblance in the things and not the points of difference.' This means that one thing may be substituted for another like to it, provided that their likeness really extends to the point in question, which can often only be ascertained with more or less probability by inductive inquiry. He adds, that the expression of the agreement must consist of a qualified judgment of identity, or a proposition of the form U, by which symbol he indicates a proposition denoted in this tract by the expression $A = B$. This exactly agrees with my view of the matter.

61. The principle of substitution of similars seems to throw a clear light upon the infinite importance of classification. For classification consists in arranging things, either in the mind or in cabinets of specimens, according to their resemblances, and the best classification is that which exhibits the most numerous and extensive resemblances. The pur-

[1] *Laws of Thought*, fifth ed. p. 251.

pose and effect of such arrangement evidently is, that we may apply to all members of a class whatever we know of any member, *so far as it is a member*. All the members of a class are mutual substitutes for each other as regards their common characteristics, and a natural classification is that which gives the greatest probability that characters as yet unexamined will exhibit agreements corresponding to those which are examined. Classification is thus the infinitely useful mode of multiplying knowledge, by rendering knowledge of particulars as general as possible, or of indicating the greatest possible number of substitutions which may give rise to acts of inference.

62. I need hardly point out that not only in our reasonings, but in our acts in common life, we observe the principle of similarity. Any new kind of action or work is performed with doubt and difficulty, because we have no knowledge derived from a similar case to guide us. But no sooner has the work been performed once or twice with success than much of the difficulty vanishes, because we have acquired all the knowledge which will guide us in similar cases. Mankind, too, have an instinctive respect for precedents, feeling that, however we act in one particular case, we ought to act similarly in all similar cases, until strong reason or necessity obliges us to make a new precedent. The whole practice of law in English courts, if not in all others, consists in deciding all new causes according to the rule established in the most nearly similar former causes, provided any can be found sufficiently similar. No ruler, too, but an absolute tyrant can perform any public act but under the responsibility of being called upon to perform a similar act, or make a similar concession, in similar circumstances.

63. At the present day, for instance, the Government is called upon to take charge of the telegraphs and railways, because great benefit has resulted from their management of the post-office. It is implied in this demand that the telegraphs and railways resemble or are even identical with the post-office, in those points which render Government control beneficial, and the public mind inevitably leaps from one thing to anything which appears similar. The whole question turns, of course, upon the degree and particular nature of the similarity. Granting that there is sufficient analogy between the telegraph and the post-office to render the Government purchase of the former desirable, we must not favour so gigantic an enterprise as the purchase of the railways until it is clearly made out that their successful management depends upon principles of economy exactly similar to the case of the post-office.

64. The great immediate question of the day is the Disestablishment of the Irish Church. The opponents of the measure argue against it by the indirect argument, that if the Irish Church ought to be disestablished, so ought the English Church; but as this ought *not*, neither ought the Irish Church. They are answered by pointing out that the Irish and English Churches are not *similarly situated*; the one possesses the sympathy of the great body of the people, and the other does not. This is an all-important point, which prevents our applying to one what we apply to the other. But on either side it is unconsciously, if not expressly, allowed that similars must be similarly treated. Almost the whole of our difficulties in the government of Ireland arise from the different national characters of the Irish and English, which renders laws and institutions suited to the one inapplicable to the other. Yet such is the tendency of indiscriminating public opinion to run in the groove of similarity, that it requires a bold legislator to repeal laws for Ireland which it is not intended or desired to repeal for England.

65. Before closing, I should notice that at some period in the obscurity of the Middle Ages an attempt seems to have been made to assimilate in some degree the logical and mathematical sciences, by inventing a logical canon analogous to the first axiom of Euclid. Between the *dictum de omni et nullo* of Aristotle, which had so long been esteemed the primary and perfect rule of reason, and the

THE TRUE PRINCIPLE OF REASONING

axiom concerning equal quantities, there was no apparent similarity. Logicians accordingly adopted a syllogistic canon which seems closely analogous to the axiom in question, and which was thus stated in the text-book of Aldrich—

Quæ conveniunt in uno aliquo eodemque tertio, ea conveniunt inter se.

This was supplemented by a corresponding canon concerning terms which disagree—

Quorum unum convenit, alterum differt uni et eidem tertio, ea differunt inter se.

The excessive subtlety of logical writers of past centuries even led them to invent six separate canons to express the principle which seems to be sufficiently embodied in our one rule. Whately considers two of these canons to be a sufficient rule of reason, which he thus translates—

If two terms agree with one and the same third, they agree with each other; and

If one term agrees, and another disagrees, with one and the same third, these two disagree with each other.

'No categorical syllogism can be faulty which does not violate these canons: none correct which does.'[1]

66. Though Wallis spoke of these canons as an innovation in his day, Mr. Mansel has traced them back to the time of Rodolphus Agricola.[2] They were well known to Lord Bacon, for he appears to have been greatly struck with the apparent analogy between these canons and the axioms of mathematicians, and he introduces it as an instance of conformity or analogy in his *Novum Organum*[3] in the following passage:—

---

[1] Whately, *Elements of Logic*, Book II, chap iii, sec. 2.
[2] Born 1442; his logical work, *De Inventione Dialecticæ*, was printed at Louvain in 1516.
[3] Book II, Aphorism 27.

Postulatum mathematicum, ut quæ eidem tertio æqualia sunt, etiam inter se sint æqualia, conforme est cum fabrica syllogismi in logica: qui unit ea quæ conveniunt in medio.

67. It is a truly curious fact in the history of Logic, that these canons should so long have been adopted, and yet that the only form of proposition to which they correctly apply should have been almost wholly ignored until the present century.

It is only when applied to propositions of the form $A = B$ that these canons prevent us from falling into error, but when used with the propositions of the old Aristotelian system they allow the free commission of fallacies of undistributed middle. It has been well pointed out by Mr. Mansel,[1] that 'these canons are an attempt to reduce all the three figures of syllogism directly to a single principle; the *dictum de omni et nullo* of Aristotle, which was universally adopted by the scholastic logicians, being directly applicable to the first figure only. This reduction, so long as the predicate of propositions has no expressed quantity, is illegitimate; the terms not being equal, but contained one within another, as is denoted by the names *major* and *minor*. Hence, as applied to the first figure, the word *conveniunt* has to express, at one and the same time, the relation of a greater to a less, and of a less to a greater, —of a predicate to a subject, and of a subject to a predicate.'

Thus in the syllogism of the mood *Barbara*,

> Metals are elements,
> Iron is a metal,
> Iron, therefore, is an element,

the terms *elements* and *iron* are both said to agree with *metals*, the third common term, although *elements* is a wider term, and *iron* a much narrower one, than *metals*. Nothing can be more unscientific and fallacious than such an applica-

[1] *Artis Logicæ Rudimenta*, p. 65.

tion of the same word in two distinct meanings. And if we avoid this fallacy by taking the meaning of the word *agreement* in the same manner in each premise, we fall into the fallacy of undistributed middle. Thus

> *Metals are elements,*
> *Oxygen is an element,*
> *Oxygen, therefore, is a metal,*

would conform precisely to the canon, because *oxygen* agrees with *element* exactly in the same sense in which *metals* agree with *elements*, and yet the result is an untrue and fallacious conclusion. Doubtless this absurdity may be explained away by pointing out that *metals* and *oxygen* do not really agree with the same part of the class *elements*, so that there is no really common third term; but the so-called supreme canon of syllogism is unable to indicate when this is the case and when it is not. Other rules have to be assumed in order to overrule the supreme rule, and these involve the principle of quantification, because they depend upon the inquiry as to what parts of the middle term are *identical* respectively with the major and minor terms. Yet for centuries logicians failed to acknowledge that identity is at the bottom of the question.

68. To sum up, we may say that the logicians attempted to reconcile logical with mathematical forms of reasoning, by assuming a canon which is true when applied to quantified propositions; but, as they applied the canon to unquantified propositions, they failed in producing anything but a fallacious appearance of conformity. In the present century logicians have abundantly recognised the importance of quantifying the proposition; but they have either adhered to the old form of canon, or they have omitted altogether to inquire into the axioms which must be adopted as the groundwork of the reasoning process. I have long felt persuaded of the truth enounced by that most clear thinker, Condillac—that 'équations, propositions, jugemens, sont au fond la même chose, et que, par conséquent, on raisonne de la même manière dans toutes les sciences;'[1] and it has been my endeavour at once to transform the proposition into the equation, and to employ it with an axiom of adequate simplicity and generality, not spoiling good new material with old tools.

69. I write this tract under the discouraging feeling that the public is little inclined to favour or to inquire into the value of anything of an abstract nature. There are numberless scientific journals and many learned societies, and they readily welcome the minutest details concerning a rare mineral, or an undescribed species, the newest scientific toy, or the latest observations concerning a change in the weather. All these things are in public favour because they come under the head of physical science. Mathematicians, again, are in favour because they help the physical philosophers: accordingly the most incomprehensible speculations concerning a quintic, or a resolvent, or a new theory of groups, are readily (and deservedly) printed, although not a score of men in England can understand them. But Logic is under the *ban* of metaphysics. It is falsely supposed to lead to no *useful works*—to be mere speculation; and, accordingly, there is no journal, and no society whatever, devoted to its study. Hardly can a paper on a logical subject be edged into the Proceedings of any learned society except under false pretences. This state of things is doubtless due to an excessive reaction against the former pre-eminence of logical studies. Bacon, in protesting against the absurdities of the scholastic logicians, and the deference paid to an ancient author, placed himself at the head of this reaction. Were he living now, he would probably see that the slow pendulum of public opinion has swung to the opposite extreme, and would employ his great intellect in showing how absurd it is to cultivate the branches of the tree of knowledge, and neglect the root—which root is undoubtedly to be found in a true comprehension of logical method.

[1] *La Logique: Œuvres de Condillac*, vol. xxii, p. 173.

APPENDIX

DESCRIPTION OF THE LOGICAL ABACUS

ALTHOUGH a brief account of the abacus is given in the text (p. 117), it seems desirable to add a more minute description, which, in connection with the drawings placed in front of the title-page, will enable copies of the abacus to be made with ease. The contrivance is of so simple a character, that an instrument-maker, or even an ordinary cabinet-maker, would probably be able to construct it from the figures and description.

The abacus consists, in the first place, of an ordinary black board of deal wood, such as is used in schools or lecture-rooms. This board should be about $3\frac{1}{2}$ feet square, and must have four ledges (1, 1) of wood, about 1 inch deep and $\frac{1}{3}$ inch thick, fixed across it at equal and parallel distances, a space of about 15 inches being left at the upper part of the board. The ledges may be made to extend quite across the width of the board, and they should be painted, like the board, of a dull black colour. When in use, the board is supported on a suitable stand in a slightly inclined position, as shown by the side view (6) in the figure, so that the slips of wood placed upon the ledges, as at (6, 7) will stand securely.

It is convenient to have altogether four sets of lettered slips, namely :—

    4 of the size shown at (8) ... $3\frac{1}{4}$ inches long.
    8 ,, ,, (10) ... $4\frac{1}{2}$,,
   16 ,, ,, (12) ... $5\frac{1}{2}$,,
   32 ,, ,, (14) and (16)—$6\frac{1}{2}$,,

At (9, 11, 13, 15, and 17) are shown side views of the same slips. They are made of the best baywood, $\frac{1}{8}$ inch thick, 1 inch broad, and of the lengths stated above, so as to give a surface of 1 square inch for each letter. Each wooden slip is marked with a different combination of letters, printed upon white paper and pasted on the face of the slip.[1] The nature of the combinations will be readily gathered from pp. 115, 116, and 117, and a set of sixteen of the slips is shown in the figure at (7), resting upon one of the ledges in the usual manner.

In the face of each wooden slip are fixed pins of thin brass or steel wire, projecting from the wood about $\frac{1}{6}$ inch in an inclined direction. Every slip has a pin near to its upper end, as at (18), but the positions of the other pins are varied according to the combination of letters represented on the slip. Each large capital or positive letter is furnished with a pin in the upper part of the space allotted to it, as at (19, 20, 21, etc.), while each small or negative letter has a pin in the lower part of its space, as at (22, 23, 24, etc.) At the lower end of each slip and at the front is fixed, as at (9, 11, 13, 15, 17), a thick square piece of sheet lead, weighing from $\frac{1}{2}$ oz. to 1 oz., so adjusted that each slip will hang in stable equilibrium and in an upright position when lifted by any of the pins. The lead may be covered at the front side with white paper.

The only other requisite is a flat straight-edge or ruler of hard wood about 16 inches long, $1\frac{1}{2}$ inch wide, and $\frac{1}{6}$ inch thick. It is shown in the figure at (2), and an enlarged section at (3), where the sharp edge will be seen to be strengthened with a slip of brass plate (4). It will be desirable to have a box made to hold all the slips in proper order, arranged in trays, so that any set may readily be taken out by the aid of the straight-edge, inserted under the row of top pins.

In using the abacus, one or other of the series of slips is taken out, according as the logical problem to be solved contains more or less terms. If there be only two terms, the set of four is used; if three, the set of eight; if four, the set of sixteen; and if five, the set of thirty-two slips. Thus the syllogism *Barbara* would require the set of eight slips (see p. 117, etc.) These must be set side by side upon the topmost ledge, as at (6): the order in which they are placed is not of any essential importance, but it is generally convenient for the sake of clearness that every positive combination should be placed on the left of the corresponding negative, and that the order shown at (7), and at pp. 115, 116, and 117, should be as much as possible maintained. When a series of the slips is resting on one of the ledges, it is evident that we may separate out those marked with

[1] I shall be happy to send a set of the printed letters to any person who may desire to have an abacus constructed.

A or any capital letter, by inserting the straight-edge horizontally beneath the proper row of pins, and then raising the slips and removing them to another ledge. The corresponding negative slips will be left where they were, owing to the absence of pins at the point where the straight-edge is placed. We have always the option, too, of removing either the A's or the a's, the B's or the b's, and so on. Successive movements will enable us to select any class or group out of the series : thus, if we took the series of sixteen, and removed first the a's and then the b's, we should have left the class of AB's, four in number. Dr. Boole based his logical notation upon the successive selection of classes, and it is this operation of thought which is represented in a concrete manner upon the abacus.

The examples given in the text (pp. 117-119) will partly serve to illustrate the use of the abacus, but I will minutely describe one more instance. Let the premises of a problem be—

(a) A is either B or C ; but
(β) B is D ; and
(γ) C is D.

Let it be required to answer any question concerning the character of the things A, B, C, D, under the above conditions.

(1) Take the set of sixteen slips and place them on the topmost or first ledge of the board.
(2) Remove the A's to the second ledge.
(3) Out of the A's, remove the B's back to the first ledge.
(4) Out of what remain, remove the C's back to the first ledge.
(5) What still remain are combinations contradicted by the first premise (a), and they are to be removed to the lowest ledge, and left there.
(6) The others having been joined together again on the first ledge, remove the B's to the second ledge.
(7) Out of the B's, remove the D's to the first ledge again, and
(8) Reject to the lowest ledge the B's which are not-D's, as contradictory of (β).
(9) Similarly, in treating (γ), remove the C's to the second ledge, return those which are D's, and reject the C's which are not-D's to the lowest ledge.

The combinations which have escaped rejection are all which

are possible under the conditions (α) (β) and (γ), and they will be found to be the following:—

A	A	A	a	a	a	a	a
B	B	b	B	B	b	b	b
C	c	C	C	c	C	b	c
D	D	D	D	D	D	D	d

To obtain the description of any class, we have now only to pick out that class by the straight-edge, and observe their nature. Thus, when the A's are picked out, we find that they always bring D with them; that is to say, all the A's are D's; this being the principal result of the problem. But we may also select any other class for examination. Thus the d's are represented by only one combination, which shows that what is not-D is neither C, B, nor A.

Even when the common conclusion of an argument is self-evident, it will be found instructive to work it upon the abacus, because the whole character of the argument and the conditions of the subject are then exhibited to the eye in the clearest manner; and while the abacus gives all conclusions which can be obtained in any other way, it often gives negative conclusions which cannot be detected or proved but by the indirect method (see p. 109). It also solves with certainty problems of such a degree of complexity that the mind could not comprehend them without some mechanical aid. In my previous little work on *Pure Logic* (see above, § 40, p. 110) I have given a number of examples, the working of which may be tested on the abacus, and other examples are to be found in Dr. Boole's *Laws of Thought*.

III

ON THE MECHANICAL PERFORMANCE

OF

LOGICAL INFERENCE

ON THE MECHANICAL PERFORMANCE

OF

LOGICAL INFERENCE

1. It is an interesting subject for reflection that from the earliest times mechanical assistance has been required in mental operations. The word *calculation* at once reminds us of the employment of pebbles for marking units, and it is asserted that the word ἀριθμὸς is also derived from the like notion of a pebble or material sign.[1] Even in the time of Aristotle the wide extension of the decimal system of numeration had been remarked and referred to the use of the fingers in reckoning; and there can be no doubt that the form of the most available arithmetical instrument, the human hand, has reacted upon the mind and moulded our numerical system into a form which we should not otherwise have selected as the best.

2. From early times, too, distinct mechanical instruments were devised to facilitate computation. The Greeks and Romans habitually employed the *abacus* or arithmetical board, consisting, in its most convenient form, of an oblong frame with a series of cross wires, each bearing ten sliding beads. The *abacus* thus supplied, as it were, an unlimited series of fingers, which furnished marks for successive higher units and allowed of the representation of any number.

[1] Professor De Morgan "On the word 'Αριθμὸs," *Proceedings of the Philological Society*, p. 9.

The Russians employ the abacus at the present day under the name of the *shtshob*, and the Chinese have from time immemorial made use of an almost exactly similar instrument called the *schwanpan*.

3. The introduction into Europe of the Arabic system of numeration caused the abacus to be generally superseded by a far more convenient system of written signs; but mathematicians are well aware that their science, however much it may advance, always requires a corresponding development of material symbols for relieving the memory and guiding the thoughts. Almost every step accomplished in the progress of the arts and sciences has produced some mechanical device for facilitating calculation or representing its result. I may mention astronomical clocks, mechanical globes, planetariums, slide rules, etc. The ingenious rods known as Napier's Bones, from the name of their inventor, or the Promptuarium Multiplicationis of the same celebrated mathematician,[1] are curious examples of the tendency to the use of material instruments.

4. As early as the seventeenth century we find that machinery was made to perform actual arithmetical calculation. The arithmetical machine of Pascal was constructed in the years 1642-1645, and was an invention worthy of that great genius. Into the peculiarities of the machines subsequently proposed or constructed by the Marquis of Worcester, Sir Samuel Morland, Leibnitz, Gersten, Scheutz, Donkin, and others we need not inquire; but it is worthy of notice that M. Thomas, of Colmar, has recently manufactured an arithmetical machine so perfect in construction and so moderate in cost, that it is frequently employed with profit in mercantile, engineering, and other calculations.

5. It was reserved for the profound genius of Mr. Babbage to make the greatest advance in mechanical calculation, by embodying in a machine the principles of the

[1] Rabdologiæ seu numerationis per virgulas libri duo : cum appendice de expeditissimo multiplicationis Promptuario. Quibus accessit et Arithmeticæ Localis Liber Unus. Authore et Inventore Joanne Nepero, Barone Merchistonii, etc. Lugduni, 1626.

calculus of differences. Automatic machinery thus became capable of computing the most complicated mathematical tables;[1] and in his subsequent design for an Analytical Engine Mr. Babbage has shown that material machinery is capable, in theory at least, of rivalling the labours of the most practised mathematicians in all branches of their science. Mind thus seems able to impress some of its highest attributes upon matter, and to create its own rival in the wheels and levers of an insensible machine.

6. It is highly remarkable that when we turn to the kindred science of logic we meet with no real mechanical aids or devices. Logical works abound, it is true, with metaphorical expressions implying a consciousness that our reasoning powers require such assistance, even in the most abstract operations of thought. In or before the fifteenth century the logical works of the greatest logician came to be commonly known as the *Organon* or *Instrument*, and, for several centuries, logic itself was defined as *Ars instrumentalis dirigens mentem nostram in cognitionem omnium intelligibilium*.

When Francis Bacon exposed the futility of the ancient deductive logic, he still held that the mind is helpless without some mechanical rule, and in the second aphorism of his *New Instrument* he thus strikingly asserts the need—

Nec manus nuda, nec Intellectus sibi permissus, multum valet; Instrumentis et auxiliis res perficitur; quibus opus est, non minus ad intellectum, quam ad manum. Atque ut instrumenta manus motum aut cient, aut regunt; ita et Instrumenta mentis, Intellectui aut suggerunt aut cavent.

7. In all such expressions, however, the word *Instrument* is used metaphorically to denote an invariable formula or rule of words, or system of procedure. Even when Raymond Lully put forth his futile scheme of a mechanical syllogistic, the mechanical apparatus consisted of nothing but written diagrams. It is rarely indeed that any invention is made without some anticipation being sooner or later discovered; but up to the present time I am totally unaware of even a

[1] See *Companion to the Almanack for* 1866, p. 5.

single previous attempt to devise or construct a machine which should perform the operations of logical inference;[1] and it is only I believe in the satirical writings of Swift that an allusion to an actual reasoning machine is to be found.[2]

8. The only reason which I can assign for this complete inability of logicians to devise a real logical instrument, is the great imperfection of the doctrines which they entertained. Until the present century logic has remained substantially as it was moulded by Aristotle 2200 years ago. Had the science of quantity thus remained stationary since the days of Pythagoras or Euclid, it is certain that we should not have heard of the arithmetical machine of Pascal, or the difference engine of Babbage. And I venture to look upon the logical machine which I am about to describe as equally a result and indication of a profound reform and extension of logical science accomplished within the present century by a series of English writers, of whom I may specially name Jeremy Bentham, George Bentham, Professor De Morgan, Archbishop Thomson, Sir W. Hamilton, and the late distinguished Fellow of the Royal Society, Dr. Boole. The result of their exertions has been to effect a breach in the supremacy of the Aristotelian logic, and to furnish us, as I shall hope to show by visible proof, with a system of logical deduction almost infinitely more general and powerful than anything to be found in the old writers. The ancient syllogism was incapable of mechanical performance because of its extreme incompleteness and crudeness, and it is only when we found our system upon the fundamental laws of thought themselves that we arrive at a system of deduction which can be embodied in a machine acting by simple and uniform movements.

9. To George Boole, even more than to any of the

[1] See note at the end of this paper, p. 172.

[2] In the recent *Life of Sir W. Hamilton*, by Professor Veitch, is given an account and figure of a wooden instrument employed by Sir W. Hamilton in his logical lectures to represent the comparative extent and intent of meaning of terms; but it was merely of an illustrative character, and does not seem to have been capable of performing any mechanical operations.

logicians I have named, this great advance in logical doctrine is due. In his *Mathematical Analysis of Logic* (1847), and in his most remarkable work *Of the Laws of Thought* (London, 1854), he first put forth the problem of logical science in its complete generality:—*Given certain logical premises or conditions, to determine the description of any class of objects under those conditions.* Such was the general problem of which the ancient logic had solved but a few isolated cases—the nineteen moods of the syllogism, the sorites, the dilemma, the disjunctive syllogism, and a few other forms. Boole showed incontestably that it was possible, by the aid of a system of mathematical signs, to deduce the conclusions of all these ancient modes of reasoning, and an indefinite number of other conclusions. Any conclusion, in short, that it was possible to deduce from any set of premises or conditions, however numerous and complicated, could be calculated by this method.

10. Yet Boole's achievement was rather to point out the extent of the problem and the possibility of solving it, than himself to give a clear and final solution. As readers of his logical works must be well aware, he shrouded the simplest logical processes in the mysterious operations of a mathematical calculus. The intricate trains of symbolic transformations, by which many of the examples in the *Laws of Thought* are solved, can be followed only by highly accomplished mathematical minds; and even a mathematician would fail to find any demonstrative force in a calculus which fearlessly employs unmeaning and incomprehensible symbols, and attributes a signification to them by a subsequent process of interpretation. It is surely sufficient to condemn the peculiar mathematical form of Boole's method, that if it were the true form of logical deduction, only well-trained mathematicians could ever comprehend the action of those laws of thought, on the habitual use of which our existence as superior beings depends.

11. Having made Boole's logical works a subject of study for many years past, I endeavoured to show in my work on Pure Logic [1] that the mysterious mathematical forms of Boole's logic are altogether superfluous, and that in one point of great importance, the employment of exclusive instead of unexclusive alternatives, he was deeply mistaken. Rejecting the mathematical dress and the erroneous conditions of his symbols, we arrive at a logical method of the utmost generality and simplicity. In a later work [2] I have given a more mature and clear view of the principles of this Calculus of Logic, and of the processes of reasoning in general, and to these works I must refer readers who may be interested in the speculative or theoretical views of the subject. In the present paper my sole purpose is to bring forward a visible and tangible proof that a new system of logical deduction has been attained. The logical machine which I am about to describe is no mere model illustrative of the fixed forms of the syllogism. It is an analytical engine of a very simple character, which performs a complete analysis of any logical problem impressed upon it. By merely reading down the premises or data of an argument on a key board representing the terms, conjunctions, copula, and stops of a sentence, the machine is caused to make such a comparison of those premises that it becomes capable of returning any answer which may be logically deduced from them. It is charged, as it were, with a certain amount of information which can be drawn from it again in any logical form which may be desired. The actual process of logical deduction is thus reduced to a purely mechanical form, and we arrive at a machine embodying the *Laws of Thought*, which may almost be said to fulfil in a substantial manner the vague idea of an organon or instrumental logic which has flitted during many centuries before the minds of logicians.

12. As the ordinary views of logic and the doctrine of

[1] *Pure Logic, or the Logic of Quality apart from Quantity: with Remarks on Boole's System, and on the Relation of Logic and Mathematics.* London, 1864 (Stanford).

[2] *The Substitution of Similars, the True Principle of Reasoning: derived from a Modification of Aristotle's Dictum.* London, 1869 (Macmillan).

the syllogism would give little or no assistance in comprehending the action of the machine, I find it necessary to preface the description of the machine itself with a brief and simple explanation of the principles of the indirect method of inference which is embodied in it, avoiding any reference to points of abstract or speculative interest which could not be suitably treated in the present paper.

13. Whatever be the form in which the rules of deductive logic are presented, their validity must rest ultimately upon the Fundamental Laws of Thought which develop the nature of Identity and Diversity. These laws are three in number. The first appears to give a definition of Identity by asserting that *a thing is identical with itself;* the second, known as the Law of Contradiction, states that *a thing cannot at the same time and place combine contradictory or opposite attributes;* whatever A and B may be it is certain that A cannot be both B and not B. This law, then, excludes from real, or even conceivable existence, any combination of opposite attributes.

The third law, commonly known as the Law of Excluded Middle, but which I prefer to call by the simpler title of the Law of Duality, asserts that *everything must either possess any given attribute or must not possess it.* A must either be B or not B. It enables us to predict anterior to all particular experience the alternatives which may be asserted of any object. When united, these laws give us the all-sufficient means of analysing the results of any assertion: the Law of Duality develops for us the classes of objects which may exist; the Law of Identity allows us to substitute for any name or term that which is asserted or known to be identical with it; while the Law of Contradiction directs us to exclude any class or alternative which is thus found to involve self-contradiction.

14. To illustrate this by the simplest possible instance, suppose we have given the assertion that

A metal is an element,

and it is required to arrive at the description of the class of *compound* or *not-elementary bodies* so far as affected by this assertion. The process of thought is as follows—

By the Law of Duality I develop the class *not-element* into two possible parts, those which are *metal* and those which are *not metal*, thus—

What is not element is either metal or not-metal.

The given premise, however, enables me to assert that *what is metal is element;* so that if I allowed the first of these alternatives to stand there would be a *not-element* which is yet an *element*. The law of contradiction directs me to exclude this alternative from further consideration, and there remains the inference, commonly known as the contra-positive of the premise, that

What is not element is not metal.

Though this is a case of the utmost simplicity, the process is capable of repeated application *ad infinitum,* and logical problems of any degree of complication can thus be solved by the direct use of the most fundamental Laws of Thought.

15. To take an instance involving three instead of two terms, let the premises be—

Iron is a metal .	(1)
Metal is element	(2)

We can, by the Law of Duality, develop any of these terms into four possible combinations. Thus

Iron is metal, element	(a)
or metal, not-element	(β)
or not-metal, element	(γ)
or not-metal, not-element .	(δ)

But the first premise informs us that iron is a metal, and thus excludes the combinations (γ) and (δ), while the second premise informs us that *metal* must be *element*, and thus further excludes the combination (β). It follows that iron must be described by the first alternative (*a*) only, and that it is an element, thus proving the conclusion of the syllogistic mood Barbara.

16. In employing this method of inference, it is soon found to be tedious to write out at full length in words the combinations of terms to be considered. It is much better to substitute for the words single letters, A, B, C, etc., which may stand in their place and bear in each problem a different meaning, just as x, y, z in algebra signify different quantities in different problems, and are really used as brief marks to be substituted for the full descriptions of those quantities. At the same time it is convenient to substitute for the corresponding negative terms small italic letters, *a*, *b*, *c*, etc.; thus

if   A denotes *iron*,   *a* denotes what is *not-iron*.
     B   „   *metal*,   *b*   „   „   *not-metal*.
     C   „   *element*,   *c*   „   „   *not-element*.

When these general terms are combined side by side, as in A B C, *a* B C, they denote a term or thing combining the properties of the separate terms. Thus A B C denotes *iron which is metal and element*; *a* B C denotes *metal which is element but not iron*. These letter terms A, B, C, *a*, *b*, *c*, etc. can, in short, be joined together in the manner of adjectives and nouns.

17. I must particularly insist upon the fact, however, that there is nothing peculiar or mysterious in these letter symbols. They have no force or meaning but such as they derive from the nouns and adjectives for which they stand as mere abbreviations, intended to save the labour of writing, and the want of clearness and conciseness attaching to a long clause or series of words. In the system put forth by Boole various symbols of obscure or even incomprehensible meaning were introduced; and it was implied that the inference came from operations different from those of common thought and common language. I am particularly anxious to prevent the misapprehension that the method of inference embodied in the machine is at all symbolic and dark, or differs from what the unaided human mind can perform in simple cases.

18. Great clearness and brevity are, however, gained by the use of letter terms; for if we take

$$A = \text{iron},$$
$$B = \text{metal},$$
$$C = \text{element};$$

then the premises of the problem considered are simply

| Iron is metal | . | . | . | A is B | . | . | (1) |
| Metal is element | . | . | . | B is C | . | . | (2) |

The combinations in which A may manifest itself are, according to the Laws of Thought,

A B C	.	.	.	.	(α)
A B c	.	.	.	.	(β)
A b C	.	.	.	.	(γ)
A b c	.	.	.	.	(δ)

But of these (γ) and (δ) are contradicted by (1) and (β) by (2). Hence

A is identical with A B C,

and this term, A B C, contains the full description of A or *iron* under the conditions (1) and (2).

Similarly, we may obtain the description of the term or class of things *not-element*, denoted by c. For by the Law of Duality c may be developed into its alternatives or possible combinations.

A B c	.	.	.	(β)
A b c	.	.	.	(δ)
a B c	.	.	.	(ζ)
a b c	.	.	.	(θ)

Of these (β) and (ζ) are contradicted by (2) and (δ) by (1); so that, excluding these contradictory terms, a b c alone

remains as the description or equivalent of the class *c*.
Hence what is *not-element*, is always *not-metallic* and is also
not-iron.

19. In practising this process of indirect inference upon
problems of even moderate complexity, it is found to be
tedious in consequence of the number of alternatives which
have to be written and considered time after time. Modes
of abbreviation can, however, be readily devised. In the
problem already considered it is evident that the same combination sometimes occurs over again, as in the cases of (β)
and (δ); and if we were desirous of deducing all the conclusions which could be drawn from the premises we should
find the combination (*a*) occurring in all the separate classes
A, AB, B, BC, AC. Similarly, the combination *a b* C
occurs in the classes *a*, *b*, C, *ab*, *a*C, *b*C, and it would be an
absurd loss of labour to examine again and again whether the
same combination is or is not contradicted by the premises.
It is certain that all the combinations of the terms A, B, C,
a, *b*, *c*, which are possible under the universal conditions of
thought and existence are but eight in number, as follows—

(a)	A B C
(β)	A B *c*
(γ)	A *b* C
(δ)	A *b* *c*
(ϵ)	*a* B C
(ζ)	*a* B *c*
(η)	*a* *b* C
(θ)	*a* *b* *c*.

All the classes of things which can possibly exist will be
represented by an appropriate selection from this list; B
will consist of (a), (β), (ϵ), and (ζ); C will consist of (a), (γ),
(ϵ), and (η); BC will consist of the combinations common to
these classes, as (a) and (ϵ); and so on. If we wish, then,
to effect a complete solution of a logical problem, it will
save much labour to make out in the first place the complete development of combinations, to examine each of these

in connection with the premises, to eliminate the inconsistent combinations, and afterwards to select from the remaining consistent combinations such as may form any class of which we desire the description. Performing these processes in the case of the premises (1) and (2), we find that of the eight conceivable combinations only four remain consistent with the premises, viz.—

$$\begin{array}{lll} \text{A B C} & & (a) \\ a \text{ B C} & & (\epsilon) \\ a \ b \text{ C} & & (\eta) \\ a \ b \ c & & (\theta) \end{array}$$

In this list of combinations the conditions (1) and (2) are, as it were, embodied and expressed, so that we at once learn that A according to those conditions consists of A B C only;

$$\begin{array}{lll} \text{B consists of } (a) \text{ or } (\epsilon) \\ b \quad ,, \quad (\eta) \text{ or } (\theta) \\ c \quad ,, \quad (\theta) \\ a \quad ,, \quad (\epsilon), (\eta) \text{ or } (\theta). \end{array}$$

20. It is easily seen that the solution of every problem which involves three terms A, B, C will consist in making a similar selection of consistent combinations from the same series of eight conceivable combinations. Problems involving four distinct terms would similarly require a series of sixteen conceivable combinations, and if five or six terms enter, there will be thirty-two or sixty-four of such combinations. These series of combinations appear to hold a position in logical science at least as important as that of the multiplication table in arithmetic or the coefficients of the binomial theorem in the higher parts of mathematics. I propose to call any such complete series of combinations a *Logical Abecedarium*, but the number of combinations increases so rapidly with the number of separate terms that I have not found it convenient to go beyond the sixty-four combinations of the six terms A, B, C, D, E, F and their negatives.

21. To a person who has once comprehended the

extreme significance and utility of the Logical Abecedarium, the whole indirect process of inference becomes reduced to the repetition of a few uniform operations of classification, selection, and elimination of contradictories. Logical deduction becomes, in short, a matter of routine, and the amount of labour required the only impediment to the solution of any question. I have directed much attention, therefore, to reduce the labour required, and have in previous publications described devices which partially accomplish this purpose. The Logical Slate consists of the complete Abecedarium engraved upon a common writing slate, and merely saves the labour of writing out the combinations.[1] The same purpose may be effected by having series of combinations printed ready upon separate sheets of paper, a series of proper length being selected for the solution of any problem, and the inconsistent combinations being struck out with the pen as they are discovered on examination with the premises.

22. A second step towards a mechanical logic was soon seen to be easy and desirable. The fixed order of the combinations in the written abecedarium renders it necessary to consider them separately, and to pick out by repeated acts of mental attention those which fall into any particular class. Considerable labour and risk of mistake thus arise. The Logical Abacus was devised to avoid these objections, and was constructed by placing the combinations of the *abecedarium* upon separate movable slips of wood, which can then be easily classified, selected and arranged according to the conditions of the problem. The construction and use of this Abacus have, however, been sufficiently described both in the *Proceedings of the Manchester Literary and Philosophical Society* for 3d April 1866, and more fully in my recently published work, called *The Substitution of Similars*, which contains a figure of the Abacus. I will only remark, therefore, that while the logical slate or printed abecedarium is convenient for the private study of logical problems, the

[1] See *Pure Logic*, p. 68.

abacus is peculiarly adapted for the logical class-room. By its use the operations of classification and selection, on which Boole's logic, and in fact any logic must be founded, can be represented, and the clearest possible solution of any question can be shown to a class of students, each step in the solution being made distinctly apparent.

23. In proceeding to explain how the process of logical deduction by the use of the abecedarium can be reduced to a purely mechanical form, I must first point out that certain simple acts of classification are alone required for the purpose. If we take the eight conceivable combinations of the terms A, B, C, and compare them with a proposition of the form

$$A \text{ is } B \quad . \quad . \quad . \quad . \quad (1)$$

we find that the combinations fall apart into three distinct groups, which may be thus indicated—

Excluded combinations $\quad . \quad . \quad . \quad \begin{cases} a & a & a & a \\ B & B & b & b \\ C & c & C & c \end{cases}$

Included combinations consistent with premise (1) $\begin{cases} A & A \\ B & B \\ C & c \end{cases}$

Included combinations inconsistent with premise (1) $\quad . \quad . \quad \begin{cases} A & A \\ b & b \\ C & c \end{cases}$

The highest group contains those combinations which are all a's, and on account of the absence of A are unaffected by the statement that A's are B's; they are thus *excluded* from the sphere of meaning of the premise, and their consistency with truth cannot be affected by that premise. The middle group contains A-combinations, included within the meaning of the premise, but which also are B-combinations, and therefore comply with the condition expressed in the premise. The lowest group consists of A-combinations also, but such as are distinguished by the absence of B, and which are therefore inconsistent with the premise requiring that where A is, there B shall be likewise. This analysis would evidently be effected most simply by placing the eight combinations of the abecedarium in the middle rank, raising the a's into a higher rank, and then lowering such b's as remain in the middle rank into a lower rank. But as we only require in the solution of a problem to eliminate the inconsistent combinations, we must unite again the two upper ranks, and we then have

Combinations consistent with the premise (1) $\begin{cases} A & A & & a & a & a & a \\ B & B & & B & B & b & b \\ C & c & & C & c & C & c \end{cases}$

Combinations inconsistent with the premise (1) . . $\begin{cases} A & A \\ b & b \\ C & c \end{cases}$

24. Supposing we now introduce the second premise,

$$\text{B is C} \qquad . \qquad . \qquad . \qquad (2)$$

the operations will be exactly similar, with the exception that certain combinations have already been eliminated from the abecedarium by the first premise. These contradicted combinations may or may not be consistent with the second premise, but in any case they cannot be readmitted. Whatever is inconsistent with any one condition, is to be deemed inconsistent throughout the problem. Hence the analysis effected by the second premise may be thus represented—

Combinations consistent with (1) $\begin{cases} \text{Combinations excluded from (2)} \quad . \quad \begin{cases} a & a \\ b & b \\ C & c \end{cases} \\ \text{Combinations included and consistent with (2)} \quad . \quad . \quad \begin{cases} A & a \\ B & B \\ C & C \end{cases} \\ \text{Combinations inconsistent with (2)} \quad . \quad \begin{cases} A & a \\ B & B \\ c & c \end{cases} \end{cases}$

Combinations inconsistent with (1) $\begin{cases} A & A \\ b & b \\ C & c \end{cases}$

To effect the above classification, we first move down to a lower rank the combinations inconsistent with (1); we then raise the b's, and out of the remaining B's lower the c's. But as all our operations are directed only to distinguish the consistent and inconsistent combinations, we now join the highest to the second rank, and the third to the lowest, as follows—

Combinations consistent with (1) and (2) . . . $\begin{cases} A & & a & a & a \\ B & & B & b & b \\ C & & C & C & c \end{cases}$

Combinations inconsistent with (1) or (2), or both . . $\begin{cases} A & A & A & a \\ B & b & b & B \\ c & C & c & c \end{cases}$

25. The problem is now solved, and it only remains to put any question we may desire. Thus if we want the description of the class A, we may raise out of the consistent combinations such as are a's, and the sole remaining combination A B C gives the description required, agreeably to our former conclusion. To obtain the description of B, we unite the consistent combinations again and raise the b's; there will remain two combinations A B C and a B C, showing that B is always C, but that, so far as the conditions of the problem go, it may or may not be A.

26. In considering such other kinds of propositions as might occur, we meet the case where two or more terms are combined together to form the subject or predicate, as in the example

A B is C,

meaning that whatever combinations contain both A and B, ought also to contain C. This case presents no difficulty; and to obtain the included combinations it is only necessary to raise out of the whole series of combinations the a's *and* b's, simultaneously or successively. The result, in whatever way we do it, is as follows—

Excluded combinations $\begin{cases} A & A & a & a & a & a \\ b & b & B & B & b & b \\ C & c & C & c & C & c \end{cases}$

Included combinations $\begin{cases} A & A \\ B & B \\ C & c \end{cases}$

We may then remove such of the included combinations, *i.e.* A B *c* only, as may be inconsistent with the premise, and proceed as before.

27. Had the predicate instead of the subject contained two terms as in

A is B C,

we should have required to raise the *a*'s and then lower the *b*'s and the *c*'s, in an exactly similar manner.

28. The only further complication to be considered arises from the occurrence of the disjunctive conjunction *or* in the subject or predicate, as in the case

A is B or C (or both).

To investigate the proper mode of treating this condition, we may take the same series of eight conceivable combinations and raise those containing *a*, in order to separate the excluded combinations. But it is not now sufficient simply to lower such of the included combinations as contain *b*, and condemn these as inconsistent with the premise. For though these combinations do not contain B they may contain C, and may require to be admitted as consistent on account of the second alternative of the predicate. While the AB's are certainly to be admitted, the A*b*'s must be subjected to a new process of selection. Now the simplest mode of preparing for this new selection is to join the AB's to the *a*'s or excluded combinations, to move up the A*b*'s into the place last occupied by the A's, to lower such of the A*b*'s as do not contain C. The result will then be as follows—

Excluded combinations and included combinations consistent with 1st alternative $\begin{cases} A & A \\ B & B \\ C & c \end{cases}$ $\begin{matrix} a & a & a & a \\ B & B & b & b \\ C & c & C & c \end{matrix}$

Included combination inconsistent with 1st but consistent with 2d alternative . . . . $\begin{cases} A \\ b \\ C \end{cases}$

Included combination inconsistent with both alternatives . . . $\begin{cases} A \\ b \\ c \end{cases}$

It is only the lowest rank of combinations, in this case containing only A b c, which is inconsistent with the premise as a whole, and which is therefore to be condemned as contradictory; and if we join the two higher ranks we have effected the requisite analysis.

29. It will be apparent that should the subject of the premise contain a disjunctive conjunction, as in

A or B is C,

a similar series of operations would have to be performed. We must not merely raise the a's and treat them as excluded combinations, but must return them to undergo a new sifting, whereby the aB's will be recognised as included in the meaning of the subject, and only the ab's will be treated as excluded. This analysis effected, the remaining operations are exactly as before.

30. The reader will perhaps have remarked that in the case of none of the premises considered has it been requisite to separate the combinations of the abecedarium into more than four groups or ranks, and it may be added that all problems involving simple logical relations only have been sufficiently represented by the examples used. The task of constructing a mechanical logic is thus reduced to that of classifying a series of wooden rods representing the conceivable combinations of the abecedarium into certain definite groups distinguished by their positions, and providing such mechanical arrangements, that wherever a letter term occurs

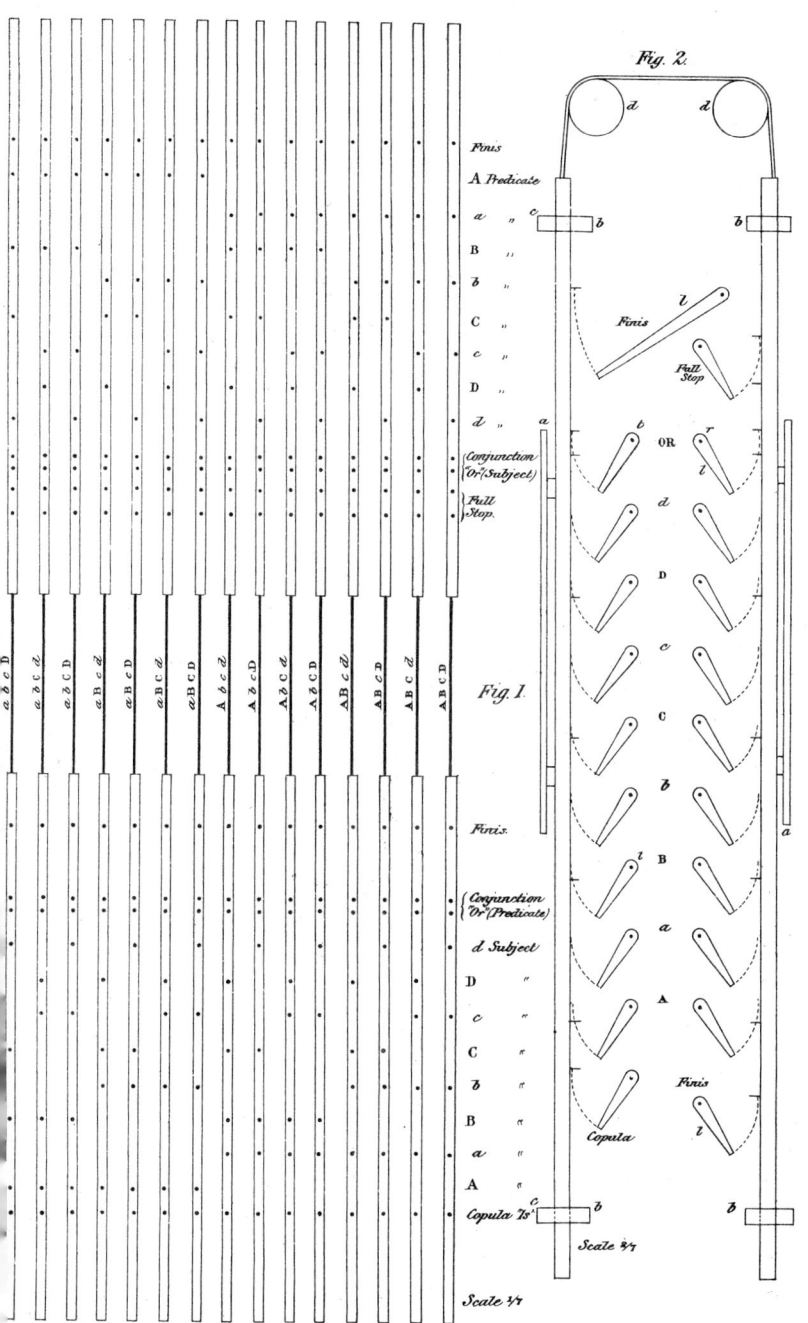

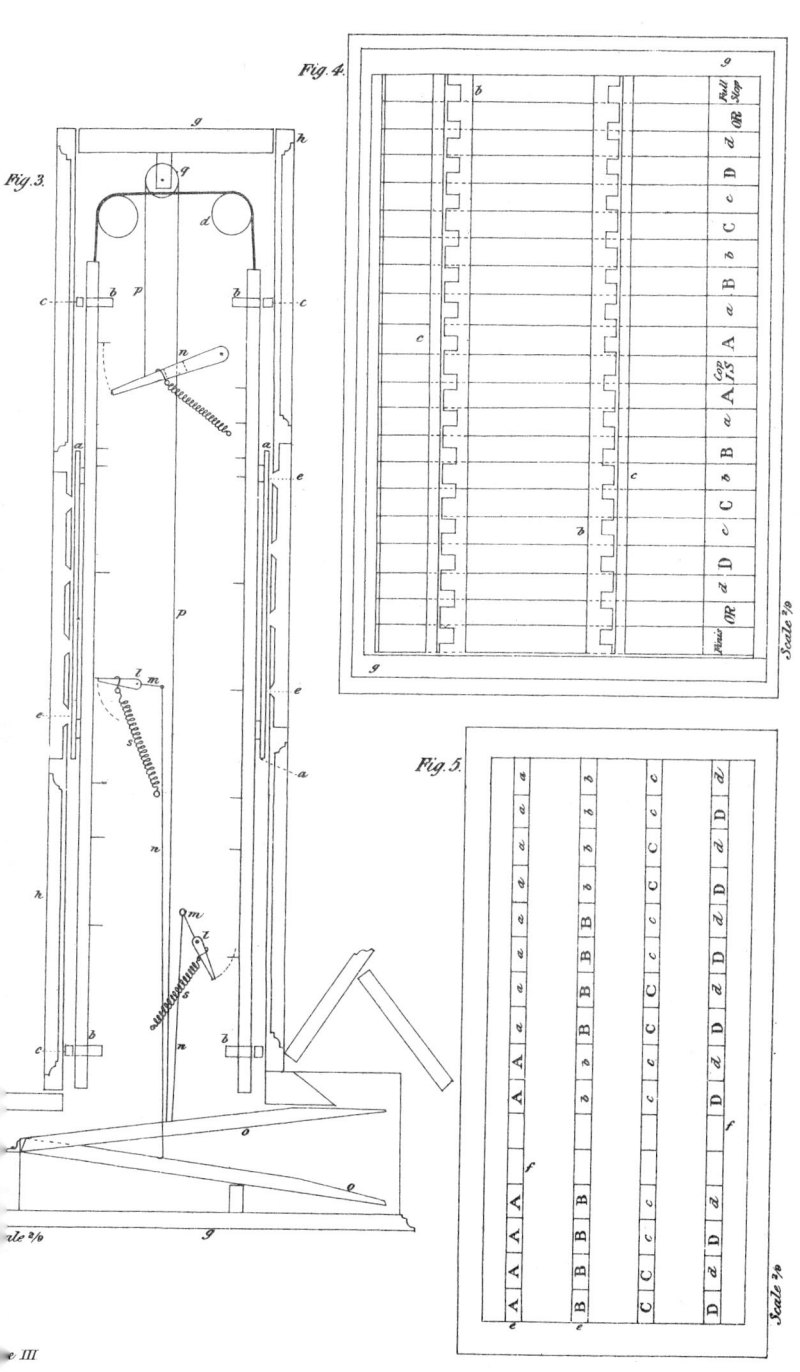

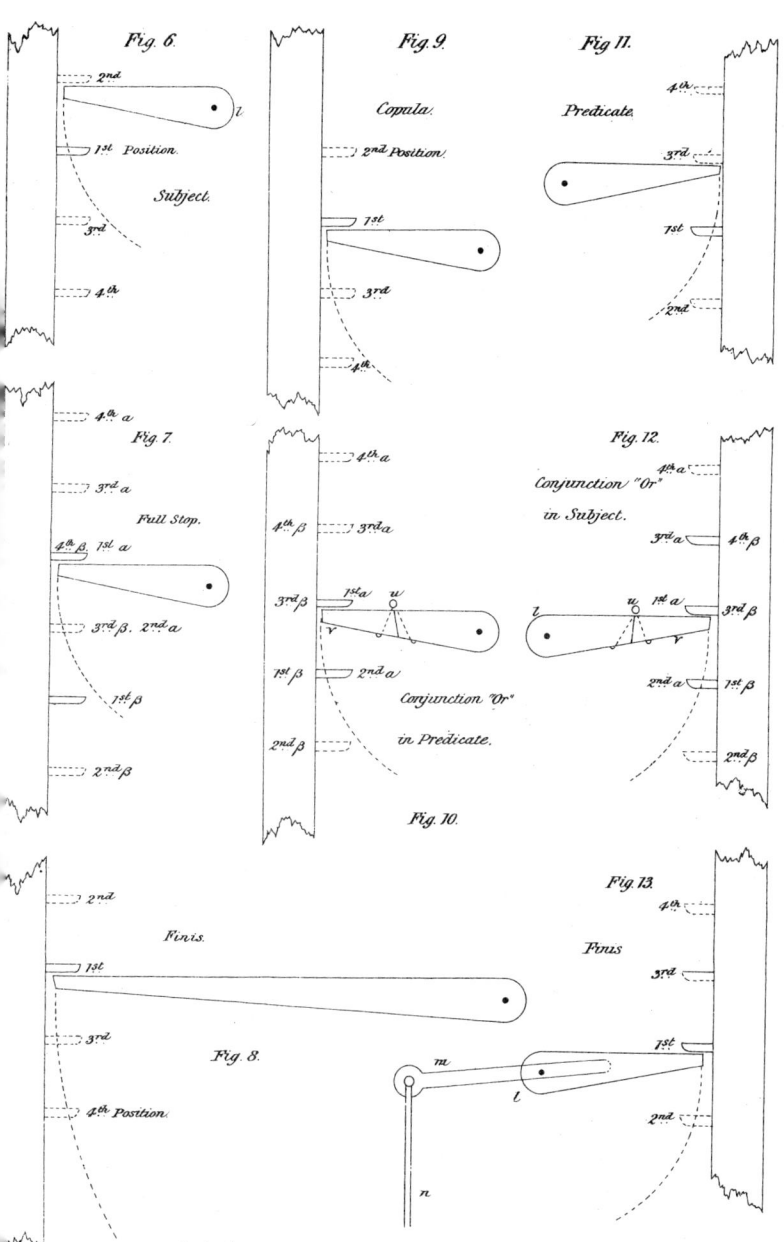

in the subject or predicate of a proposition, or a conjunction copula or stop intervenes, the pressure of a corresponding lever or key shall execute systematically the required movements of the combinations.

31. The principles upon which the logical machine is based will now be apparent to the reader; and as the construction of the machine involves no mechanical difficulties of any importance, it only remains for me to give as clear a description of its component parts and movements as their somewhat perplexing character admits of.

32. The Machine which has been actually finished is adapted to the solution of any problems not involving more than four distinct positive terms, indicated by A, B, C, D, with, of course, their corresponding negatives, a, b, c, d. The requisite combinations of the abecedarium are, therefore, sixteen in number (§ 20, p. 174), and each combination is represented by a pair of square rods of baywood (Plate II, fig. 1), united by a short piece of cord and slung over two round horizontal bars of wood (d, d, figs. 2 and 3), so as to balance each other and to slide freely and perpendicularly in wooden collars (b, b, figs. 2, 3, and 4) closed by plain wooden bars (c, c). To each rod is attached a thin piece of baywood, $8\frac{1}{2}$ inches long and 1 inch wide (a, a, figs. 2 and 3), bearing the letters of the combination represented. Each letter occupies a space of $\frac{1}{2}$ inch in height, but is separated from the adjoining letter by a blank space of white paper $1\frac{1}{2}$ inch long. Both at the front and back of the machine are pierced four horizontal slits, $1\frac{1}{2}$ inch apart, extending the whole width of the case, and $\frac{1}{2}$ inch in height, so placed that when the rods are in their normal position each letter shall be visible through a slit. The machine thus exhibits on its two sides, when the rods are in a certain position, the combinations of the abecedarium as shown in fig. 5; but should any of the rods be moved upwards or downwards through a certain limited distance the letters will become invisible as at f, f (Plate III, fig. 5).

33. Externally the machine consists of a framework,

seen in perpendicular section in fig. 3 (g, g), and in horizontal section in fig. 4 (g, g), which serves at once to support and contain the moving parts. It is closed at the front and back by large doors (h, h, fig. 3), in the middle panels of which are pierced the slits rendering the letters of the abecedarium visible.

34. The rods are moved upwards, and the opposite rods of each pair are thus caused to fall downwards, by a series of long flat levers seen in section at l, l, l (figs. 2, 3, 6 . . . 13). These levers revolve on pivots inserted in the thicker part, and move in sockets attached to the inner side of the framework. Brass arms (m, m, figs. 3 and 13), connected by copper wires (n, n) with the keys of the machine (o, o), actuate the levers, which are caused to return, when the key is released, by spiral brass springs (s, s).

35. The levers communicate motion to the rods by means of brass pins fixed in the inner side of the rods (fig. 2). As it is upon the peculiar arrangement of these pins that the whole action of the machine depends, the position of each of the 272 pins is shown by a dot in fig. 1, in which are also indicated the function of each pin and the combination represented by each pair of rods. It is seen that certain pins are placed uniformly in all the adjoining rods, as in the rows opposite the words *Finis, Conjunction, Copula, Full Stop*. These may be called *operation pins*, and must be distinguished from the *letter pins*, representing the terms of the combination, and varied in each pair of rods to correspond with the letters of the abecedarium. On examining fig. 1, it will be apparent that the pins are distributed in a negative manner; that is to say, it is the absence of a pin in the space A, and its presence in the space a, which constitutes the rod a representative of the term A. The rods belonging to the combination A b C d, for instance, have pins in the spaces belonging to the letters a, B, c, D.

36. The key board of the instrument is shown in fig. 4, where are seen two sets of term or letter keys, marked A, a, B, b, C, c, D, d, separated by a key marked COPULA—

Is. The letter keys on the left belong to the subject of a proposition, those on the right to the predicate, and on either side just beyond the letter keys is a *Conjunction* key, appropriated to the disjunctive conjunction OR, according as it occurs in the subject or predicate. The last key on the right hand is marked FULL STOP, and is to be pressed at the end of each proposition, where the full stop is properly placed. On the extreme left, lastly, is a key marked FINIS, which is used to terminate one problem and prepare the machine for a new one.

37. In order to gain a clear comprehension of the action of these keys, we must now turn to fig. 2, where all the levers are shown in position, only three of them being inserted in fig. 3, and to figs. 6-13 (Plate IV), which represent, in the full natural size, the relative positions of each kind of lever with regard to the pins in every possible position of the rods.

If the subject key A be pressed it actuates the lever A at the back of the machine; and supposing all the rods to be in their proper initial positions, it moves upwards, as in fig. 6, all the back a rods through exactly $\frac{1}{2}$ inch, the front rods connected with them of course falling through $\frac{1}{2}$ inch. All the a combinations are thus caused to disappear from the abecedarium; but as the A rods have no pins opposite to the A lever, they will remain unmoved, and continue visible. Thus the pressure of the A key effects the selection of the class A of the conceivable combinations. Each subject letter key similarly acts upon a lever at the back; and should several of them be pressed, either simultaneously or in succession, the combinations containing the corresponding letters will be selected.

38. Each predicate letter key is connected with a lever in the front of the machine, and when pressed the effect is exactly the same as that of a subject key, but in the opposite direction (fig. 11). If the B predicate key be pressed it raises through $\frac{1}{2}$ inch all the front rods which happen to have corresponding pins, and to be in the initial position.

The back rods will at the same time fall, and the combinations containing b will disappear from the abecedarium, but in the opposite direction.

39. It is now necessary to explain that each rod has four possible positions fully indicated in the figs. 6-13. The first of these positions is the neutral or initial position, in which the letters are visible in the *abecedarium*, and the letter pins are opposite letter levers so as to be acted upon by them. The second position is that into which a rod is thrown by a subject key; the third position lies in the opposite direction, and is that into which a rod is thrown by a predicate key. The fourth position lies $\frac{1}{2}$ inch beyond the third. The four positions evidently correspond to the four classes into which combinations were classified in the previous part of the paper as follows—

Second position.—Combinations excluded from the sphere of the premise.

First position.—Combinations included, but consistent with the premise.

Third position.—Temporary position of combinations contradicted by the premise: also temporary position of combinations excluded from some of the alternatives of a disjunctive predicate.

Fourth position.—Final position of contradictory or inconsistent combinations.

40. Let us now follow out the motions produced by impressing the simple proposition

$$A \text{ is } B$$

upon the machine, all the rods being at first in the initial or first position. The keys to be pressed in succession are—

>First. The subject key, A.
>Second. The copula key.
>Third. The predicate key, B.
>Fourth. The full-stop key.

The subject key A has the effect of throwing all the a

rods from the first into the second position, the back rods rising and the front rods falling $\frac{1}{2}$ inch.

The copula key will in this case have no effect, for, as seen in fig. 9 (Plate IV), it acts only on rods in the third position, of which there are at present none.

The predicate key (fig. 11) does not act upon such of the rods (those marked a) as are in the second position, but it acts upon those in the first position, provided they have pins opposite the lever. The effect thus far will be that the a rods are in the second position, the Ab rods in the third, while the AB rods remain undisturbed in the first position. An analysis has been effected exactly similar to that explained above (§ 23, p. 176).

41. The full-stop key being now pressed has a double effect. It acts only on a single lever at the front of the machine (figs. 2 and 7), but the front rods all have in the space opposite to the lever two pins 1 inch apart (fig. 1). These pins we may distinguish as the a and β pins, the a pin being the uppermost. While a rod is in the first position the lever passes between the pins and has no effect; but if the rod be lowered $\frac{1}{2}$ inch into the second position, the lever will cause the rod to return to the first position by means of the a pin; but if the rod be raised into the third position, the β pin will come into gear, and the rod will be pushed $\frac{1}{2}$ inch further into the fourth position. Now in the case we are examining, the AB's are in the first position and will so remain; the a's are in the second and will return to the first, the Ab's are in the third, and will therefore proceed onwards to the fourth. The reader will now see that we have effected the classification of the combinations as required into those consistent with the premise A is B, whether they be included or not in the term A, and those contradicted by the premise which have been ejected into the fourth position. An examination of the figures 6-13 will show that only one lever (fig. 8) moved by the finis key affects rods in the fourth position, so that any combination rod once condemned as contradictory so remains until the close of the problem, and its letters are no more seen upon the abecedarium.

42. Any other proposition, for instance, B is C, can now be impressed on the keys, and the effects are exactly similar, except that the Ab combinations are out of reach of the levers. The B subject key throws the b's into the second, the C predicate key throws the Bc's into the third, and the full stop throws the latter into the fourth, where they join the Ab's already in that place of exclusion, while the remainder all return to the first position.

The combinations now visible in the abecedarium will be as follows—

A	A		a	a		a	a	a	a
B	B		B	B		b	b	b	b
C	C		C	C		C	C	c	c
D	d		D	d		D	d	D	d

They correspond exactly to those previously obtained from the same premises (see § 24), except that each combination of A, B, C, a, b, c is repeated with D and d. If we now want a description of the term A, we press the subject key A, and all disappear except

A B C D,   A B C d,

which contain the information that A is always associated with B and with C, but that it may appear with D or without D, the conditions of the problem having given us no information on this point. The series of consistent combinations is restored at any time by the full-stop key, the contradictory ones remaining excluded.

43. Any other subject key or succession of subject keys being pressed gives us the description of the corresponding terms. Thus the key c gives us two combinations, $a\ b\ c\ $D, $a\ b\ c\ d$, informing us that the absence of C is always accompanied by the absence of A and B. Of b we get the description

$$\begin{array}{cccc} a & a & a & a \\ b & b & b & b \\ C & C & c & c \\ D & d & D & d \end{array}$$

whence we learn that the absence of B always causes the absence of A, but that C and D are indifferently absent or present.

44. We can at any time add a new condition to the problem by pressing the full stop to bring the combinations as yet possible into the first position, and then impressing the new condition on the keys as before. Let this condition be

C is D.

The effect will obviously be to remove such Cd combinations as yet remain into the fourth position, leaving only five—

$$\begin{array}{ccccc} A & a & a & a & a \\ B & B & b & b & b \\ C & C & C & c & c \\ D & D & D & D & d \end{array}$$

hence we learn that A, B, and C are all D; that B, C, and D may or may not be A; that what is not D is not A, not B and not C; and so on. The conditions of this problem form what would be called a Sorites in the old logic, and we have not only obtained its conclusion A is D, but have performed a complete analysis of its conditions, and the inferences which may be drawn from those conditions.

45. The problem being supposed complete, we press the Finis key, which differs from all the others in moving two levers, one of which (fig. 13) is of the ordinary character and returns any rods which may happen to be in the second position into the first, while the other (fig. 8) has a much longer radius, is moved by a cord or flexible wire p, passing over a pulley q and through a perforation r in the flat board which forms the lever itself, in this case a lever of the second order. This broad lever sweeps the rods from the fourth position as well as any which may be in the third into the first, and together with the other lever (fig. 13) it reduces the whole of the rods to the neutral position, and renders the machine, as it were, a *tabula rasa,* upon which an entirely new set of conditions may be impressed independently of previous ones. Its office thus is to obliterate the effects of former problems.

46. When several of the letter keys on the subject side only or the predicate side only are pressed in succession, the effect is to select the combinations possessing all the letters marked on the keys. Thus if the keys A, B, C be pressed there will remain in the abecedarium only the combinations A B C D and A B C d; and if the key D be now pressed, the latter combination will disappear, and A B C D will alone remain. The effect will be exactly the same whatever the order in which the keys are pressed, and if they be pressed simultaneously there will be no difference in the result. The machine thus perfectly represents the commutative character of logical symbols which Mr. Boole has dwelt upon in pp. 29-30 of the *Laws of Thought.* What I have called the Law of Simplicity of logical symbols, expressed by the formula $AA = A$,[1] is also perfectly fulfilled in the machine; for if the same key be pressed two or more times in succession, there will be no more effect than when it is pressed once. Thus the succession of keys A A C B B A C would have merely the effect of A B C. This applies also to the predicate keys, but not of course to an alternation of subject and predicate keys.

47. To impress upon the machine the condition

A B is C D,

or *whatever combines the properties of* A *and* B *combines the properties of* C D, we strike in succession the subject keys

[1] *Pure Logic,* p. 15. Boole's *Laws of Thought,* p. 31.

A and B, the copula, the predicate keys C and D and the full stop. The subject keys throw into the second position both the a combinations and the b's; the predicate keys, out of the remaining AB's, throw the c's and d's into the third position; and the full stop completes the separation of the consistent and contradictory combinations in the usual manner.

48. It yet remains for us to consider a proposition with a disjunctive term in subject, predicate, or both members. For such propositions the conjunction keys are requisite, that adjoining the subject keys (fig. 4) for the subject, and the other for the predicate. These keys act in opposition to each other, and each is opposed, again, to its corresponding letter keys. Thus while the subject keys act on levers at the back of the machine (Plate III, fig. 3), the subject conjunction key acts on the lever r in front, while the predicate conjunction key t is at the back. These levers are shown in their full size in figs. 10 and 12, and are seen to differ from all the other levers in having the edge v moving on small wire hinges u in such a way that it can exert force upwards but not downwards. The lever can thus raise the rods; but in case it should strike a pin in returning, the edge yields and passes the pin without moving the rod. In connection with these levers each rod has two pins (figs. 1 and 2) at a distance of only $\frac{1}{2}$ inch, and the peculiar effect of these pins will be gathered from figs. 10 and 12 (Plate IV). Thus if we press in succession the predicate keys

A or B,

the key A will throw the a's into the third position. The conjunction key will now act upon the a pins of the A's and move them into the second position, and at the same time upon the β pins of the a's and return them into the first position. The key B now selects from the a's those which are b's, and puts them into the third position ready for exclusion by the full stop, which will also join to the

aB's still remaining in the first position the A's which were temporarily put out of the way in the second position. Should there be, however, another alternative, as in the term

A or B or C,

the conjunction key would be again pressed, which gives the ab's a new chance by returning them to the first, and the key C selects only the abc's for exclusion. The action would be exactly similar with a fourth alternative.

49. The subject conjunction key is similar but opposite in action. If the subject key A be pressed it throws the a's into the second position; the conjunction key then acts upon the a pin of the a's returning them to the first position, and also upon the β pin of the A's, sending them to a temporary seclusion in the third position. The key B would now select the ab's for the second position; the conjunction key again pressed would return them, and add the aB's to those in the third, and so on. The final result would be that those combinations excluded from all the alternatives would be found in the second position, while those included in one or more alternatives would be partly in the first and partly in the third positions.

In the progress of a proposition the copula key would now have to be pressed, and when the subject is a disjunctive term its action is essential. It has the effect (fig. 9) of throwing any combinations which are in the third back into the first. It thus joins together all the combinations included in one or more alternatives of the subject, and prepares them for the due action of the predicate keys.

50. It must be carefully observed that any doubly universal proposition of the form

all A's are all B's,

or, in another form of expression,

$$A = B,$$

can only be impressed upon the logical machine in the form of two ordinary propositions; thus,

$$\text{all A's are B's,}$$
and
$$\text{all B's are A's.}$$

The first of these excludes such A's as may be not-B's; the second excludes such B's as may be not-A's.

If we impress upon the keys of the machine the six propositions expressing the complete identity of A, B, C, and D, it is obvious that there would remain only the two combinations

$$\text{A B C D,}$$
$$a\ b\ c\ d,$$

the identity of the positive terms involving the identity also of their negatives.

The premise

$$\text{A or B} = \text{C or D}$$

would require to be read

$$\text{A or B is C or D,}$$
$$\text{C or D is A or B.}$$

51. To give some notion of the degree of facility with which logical problems may be solved with the machine, I will adduce the logical problem employed by Boole to illustrate the powers of his system at p. 118 of the *Laws of Thought*.

'Suppose that an analysis of the properties of a particular class of substances has led to the following general conclusions, viz.—

'1st. That wherever the properties A and B are combined, either the property C, or the property D, is present also; but they are not jointly present.

'2d. That wherever the properties B and C are combined, the properties A and D are either both present with them, or both absent.

'3d. That wherever the properties A and B are both absent, the properties C and D are both absent also; and *vice versâ*, where the properties C and D are both absent, A and B are both absent also.'

This somewhat complex problem is solved in Boole's work by a very difficult and lengthy series of eliminations, developments, and algebraic multiplications. Two or three pages are required to indicate the successive stages of the solution, and the details of the algebraic work would probably occupy many more pages. Upon the machine the problem is worked by the successive pressure of the following keys:—

1st. A, B, Copula, C, *d*, Conjunction, *c*, D, Full stop.
2d. B, C, Copula, A, D, Conjunction, *a*, *d*, Full stop.
3d. *a*, *b*, Copula, *c*, *d*, Full stop.
   c, *d*, Copula, *a*, *b*, Full stop.

There will then be found to remain in the abecedarium the following combinations:—

A B *c* D   *a* B C *d*
A *b* C D   *a* B *c* D
A *b* C *d* *a* *b* *c* *d*
A *b* *c* D

On pressing the subject key A, the A combinations printed above in the left-hand column will alone remain, and on examining them they yield the same conclusion as Boole's equation (p. 120), namely, 'Wherever the property A is present, there either C is present and B absent, or C is absent.'

Pressing the full-stop key to restore the *a* combinations, and then the keys *b*, C, we have the two combinations

A *b* C D,
A *b* C *d*,

from which we read Boole's conclusion (p. 120), 'Wherever the property C is present, and the property B absent, there the property A is present.' In a similar manner the other conclusions given by Boole in p. 129 can be drawn from the abecedarium.

52. It is to be allowed that a certain mental process of interpreting and reducing to simple terms the indications of the combinations is required, for which no mechanical provision is made in the machine as at present constructed, but an exactly similar mental process is required in the Indirect Process of Inference, as stated in my *Pure Logic*, pp. 42, 43 ; and equivalent processes are necessary in Boole's mathematical system. The machine does not therefore supersede the use of mental agency altogether, but it nevertheless supersedes it in most important steps of the process.

53. This mechanical process of inference proceeds by the continual selection and classification of the conceivable combinations into three or four groups. It should be noticed that in Boole's system the same groups are indicated by certain quasi-mathematical symbols as follows—

The coefficient $\frac{0}{0}$ indicates an excluded combination
,, $\frac{1}{1}$,, included ,,
,, $\frac{0}{1}$,, inconsistent ,,
,, $\frac{1}{0}$,, inconsistent ,,

It is exceedingly questionable whether there is any analogy at all between the significations of these symbols in mathematics and those which Boole imposed upon them in logic. In reality the symbol 1 denotes in Boole's logic inclusion of a combination under a term, and 0 exclusion. Accordingly $\frac{1}{0}$ indicates that the combination is included in the subject and not in the predicate, and is therefore inconsistent with the proposition, and $\frac{0}{1}$ indicates inclusion in the predicate and exclusion from the subject of an equational proposition or identity, from which also results inconsistency. Inclusion in both terms is indicated by $\frac{1}{1}$, and exclusion from both $\frac{0}{0}$, in which case the combination is consistent with the proposition.

54. To the reader of the preceding paper it will be evident that mechanism is capable of replacing for the most part the action of thought required in the performance of logical deduction. Having once written down the conditions or premises of an argument in a clear and logical form, we have but to press a succession of keys in the order corresponding to the terms, conjunctions, and other parts of the propositions, in order to effect a complete analysis of the argument. Mental agency is required only in interpreting correctly the grammatical structure of the premises, and in gathering from the letters of the abecedarium the purport of the reply. The intermediate process of deduction is effected in a material form. The parts of the machine embody the conditions of correct thinking; the rods are just as numerous as the Law of Duality requires in order that every conceivable union of qualities may have its representative; no rod breaks the Law of Contradiction by representing at the same time terms that are necessarily inconsistent; and it has been pointed out that the peculiar characters of logical symbols expressed in the Laws of Simplicity and Commutativeness are also observed in the action of the keys and levers. The machine is thus the embodiment of a true symbolic method or Calculus. The representative rods must be classified, selected, or rejected by the reading of a proposition in a manner exactly answering to that in which a reasoning mind should treat its ideas. At every step in the progress of a problem, therefore, the abecedarium necessarily indicates the proper condition of a mind exempt from mistake.

55. I may add a few words to deprecate the notion that I attribute much practical utility to this mechanical device. I believe, indeed, that it may be used with much advantage in the logical class-room, for which purpose it is more convenient than the logical abacus which I have already employed in this manner. The logical machine may become a powerful means of instruction at some future time by presenting to a body of students a clear and visible analysis of logical problems of any degree of complexity, and rendering each step of its solution plain. Its employment, however, in this way must for the present be restricted, or almost entirely prevented, by the predominance of the ancient Aristotelian logic, and the almost puerile character of the current logical examples.

56. The chief importance of the machine is of a purely theoretical kind. It demonstrates in a convincing manner the existence of an all-embracing system of Indirect Inference, the very existence of which was hardly suspected before the appearance of Boole's logical works. I have often deplored the fact that though these works were published in the years 1847 and 1854, the current handbooks, and even the most extensive treatises on logic, have remained wholly unaffected thereby.[1] It would be possible to search the works of two very different but leading thinkers, Mr. J. S. Mill and Sir W. Hamilton, without meeting the name of Dr. Boole, or the slightest hint of his great logical discoveries; and other eminent logicians, such as Professor De Morgan or Archbishop Thomson, barely refer to his works in a few appreciative sentences. This unfortunate neglect is partly due to the great novelty of Boole's views, which prevents them from fitting readily into the current logical doctrines. It is partly due also to the obscure, difficult, and, in many important points, the mistaken form in which Boole put forth his system; and my object will be fully accomplished should this machine be considered to demonstrate the existence and illustrate the nature of a very simple and obvious method of Indirect Inference of which Dr. Boole was substantially the discoverer,

---

[1] Professor Bain's treatise on *Logic*, which has been published since this paper was written, is an exception. In the first Part, which treats of Deductive Logic, pp. 190-207, he gives a description and review of Boole's Mathematical System; but it is significant that he omits the process of mathematical deduction where it is in the least complex, and merely quotes Boole's conclusions. Thus we have the anomalous result that in a treatise on Logical Deduction, the reader has to look elsewhere for processes which, according to Boole, must form the very basis of Deduction.

NOTE to § 7.

It has been pointed out to me by Mr. White, and has also been noticed in *Nature* (10th March 1870, vol. i. p. 487), that in the year 1851, Mr. Alfred Smee, F.R.S., the Surgeon of the Bank of England, published a work called *The Process of Thought adapted to words and language, together with a description of the Relational and Differential Machines* (Longmans), which alludes to the mechanical performance of thought.

After perusing this work, which was unknown to me when writing the paper, it cannot be doubted that Mr. Smee contemplated the representation by mechanism of *certain* mental processes. His ideas on this subject are characterised by much of the ingenuity which he is well known to have displayed in other branches of science. But it will be found on examination that his designs have no connexion with mine. His represent the mental states or operations of memory and judgment, whereas my machine performs logical inference. So far as I can ascertain from the obscure descriptions and imperfect drawings given by Mr. Smee, his Relational Machine is a kind of Mechanical Dictionary, so constructed that if one word be proposed its relations to all other words will be mechanically exhibited. The Differential Machine was to be employed for comparing ideas and ascertaining their agreement and difference. It might be roughly likened to a patent lock, the opening of which proves the agreement of the tumblers and the key.

It does not appear, again, that the machines were ever constructed, although Mr. Smee made some attempts to reduce his designs to practice. Indeed he almost allows that the Relational Machine is a purely visionary existence when he mentions that it would, if constructed, occupy an area as large as London.—10th October 1870.

IV

ON A GENERAL SYSTEM
OF
NUMERICALLY DEFINITE REASONING

ON A GENERAL SYSTEM

OF

NUMERICALLY DEFINITE REASONING

THE system of numerical reasoning described in this paper arises from the combination of arithmetical or algebraical calculation with logical reasoning. The purpose is to determine, as far as possible, the numbers of individual objects which may compose classes or groups of objects under any given logical conditions—the data consisting of those logical conditions, and the numbers of individuals in certain other related classes explained.

Only two or three previous writers have bestowed attention on this subject. Professor De Morgan is probably the first logician who pointed out that syllogistic arguments may exist in which the numbers of objects forming the several terms of the syllogism may be exactly defined, and that inference is often possible with such premises when it would not otherwise be valid. Logicians have for ages introduced notions of quantity into the syllogism; but they restricted themselves to the vague quantities *all, a part*, or *none*. Professor De Morgan enjoys the high honour of showing that definite numbers may also be the subject of syllogistic argument; and his system is fully stated in the eighth chapter of his *Formal Logic*, 'On the Numerically Definite Syllogism.'[1]

[1] See also an abstract in his *Syllabus of a Proposed System of Logic*, p. 27.

2. The late Professor Boole has also treated this subject, but under a different name, and in a very different form. His chapter on the subject[1] is entitled 'On Statistical Conditions,' by which he evidently means, 'Numerical Conditions.' It contains a most remarkable and powerful attempt to erect a general method for ascertaining the higher and lower limits of logical classes, to be employed as a subsidiary portion of his general calculus of probabilities. A paper on the same subject had been previously written by him, and entitled 'Of Propositions Numerically Definite,' but was only published after his death, by Professor De Morgan in the *Transactions of the Cambridge Philosophical Society* (vol. xi. part ii. 1868). Of these writings of Professor Boole I must say, what I have elsewhere said of other portions of his writings, that they appear in themselves perfect and almost inimitable. At the same time I must add that Boole's extraordinary power of analysis, and his perfect command of symbolic methods, usually led him to overestimate the part they should play in reasoning, and to under-estimate the value of a simple and intuitive comprehension of the subject. The very principle which he fearlessly adopts, that unintelligible symbols may give intelligible and even demonstrative results, will probably be rejected by future mathematicians, as it has been lately rejected in the strongest terms by Mr. Sandeman.[2]

3. As Mr. Boole's logical views were the basis from which I started in forming the simple but general system of logical forms explained in my *Pure Logic* in the year 1864, and, more simply still, in my *Substitution of Similars*, published in 1869, so the numerically definite system of reasoning which is here described arises from a simplification of the previous methods of De Morgan and Boole.

4. In the qualitative system of logic, which I have given

[1] *Laws of Thought*, chap. xix. p. 295. The use of the word statistical as equivalent to numerical is erroneous, although sanctioned by so high an authority as Sir J. Herschel, who applied it to the numbering of the stars. Statistical means what refers to the State or People.

[2] *Pelicotetics*, 1868, Preface, pp. ix. and x.

NUMERICALLY DEFINITE REASONING

in the works referred to, a term is taken to mean the quality or group of qualities which belong to and mark out a class of objects. Thus, the general term A stands for any group of qualities belonging to a class of objects.

Let the term, when enclosed in brackets, acquire a quantitative meaning, so as to denote the number of individuals or objects which possess those qualities. Then

(A) = number of objects possessing qualities of A, or say, for the sake of brevity, the number of A's. If, for instance,

A = character and quality of being a Member of Parliament,

(A) = number of existing Members of Parliament = 658.

5. Every logical proposition or equation now gives rise to a corresponding numerical equation. Sameness of qualities occasions sameness of numbers. Hence if

$$A = B$$

denotes the identity of the qualities of A and B, we may conclude that

$$(A) = (B).$$

It is evident that exactly those objects, and those objects only, which are comprehended under A must be comprehended under B. It follows that wherever we can draw an equation of qualities, we can draw a similar equation of numbers. Thus, from

$$A = B = C,$$

we infer

$$A = C;$$

and similarly from

$$(A) = (B) = (C),$$

meaning the number of A's and C's are equal to the number of B's, we can infer

$$(A) = (C).$$

But, curiously enough, this does not apply to negative propositions and inequalities. For if

$$A = B \smallfrown D$$

means that A is identical with B, which differs from D, it does not follow that

$$(A) = (B) \smallfrown (D).$$

Two classes of objects may differ in qualities, and yet they may agree in number. This is a point which strongly confirms me in the opinion I have already expressed, that all inference really depends upon equations, not differences;[1] and I shall therefore employ throughout this paper only equations which may be almost indifferently used in the qualitative or quantitative meaning.

6. I shall employ, as in logic, a joint term, such as A B (or A B C), to mean the class possessing all the qualities of A and B (or of A and B and C). To every positive term there corresponds a negative term, denoted by the corresponding small italic letter. Thus the negative of A is a, of B b, and so on. If, then, A means *man*, a means simply *not man*. Hence $a\,b$ will mean the combination of qualities of *not being* A and *not being* B.

7. The sign $\cdot|\cdot$ is used to stand for the disjunctive conjunction *or*, but in an unexclusive sense. Thus

$$A = B \cdot|\cdot C$$

means that whatever has the qualities of A, must have either the qualities of B or of C; but it *may have the qualities of both*. This unexclusive character of the terms and signs of logic, which creates a profound difference between my system and that of Professor Boole, prevents me from converting alternatives into numbers as they stand. It does not follow from the statement that A is either B or C, that the number of A's is equal to the number of B's added to the number of C's; for some objects, or possibly all, have been counted twice in this addition. Thus, if we say *An elector is either an elector for a borough, or for a county, or for a university*, it does not follow that the total number of

---

[1] *Substitution of Similars*, pp. 16, 17.

electors is equal to the number of borough, county, and university electors added together; for some men may be found in two or three of the classes.

8. This difficulty, however, is avoided with great ease, for we need only develop each alternative into all its possible subclasses and strike out any subclass which appears more than once, and then convert into numbers. Thus, from

$$A = B \cdot|\cdot C$$

we get

$$A = BC \cdot|\cdot Bc \cdot|\cdot BC \cdot|\cdot bC;$$

but striking out one of the terms BC as being superfluous, we have

$$A = BC \cdot|\cdot Bc \cdot|\cdot bC.$$

The alternatives are now strictly exclusive, or devoid of any common part, so that we may draw the numerical equation

$$(A) = (BC) + (Bc) + (bC).$$

Thus, if A = elector,
B = borough elector,
C = county elector,
D = university elector,

we may from the proposition,

$$A = B \cdot|\cdot C \cdot|\cdot D$$

draw the numerical equation

$$(A) = (BCD) + (BCd) + (BcD) + (Bcd) + (bCD) + (bCd) + (bcD).$$

9. The process of development employed above is the great peculiarity of Professor Boole's system of logic, and that which I have adopted. It depends upon the primary law of thought, usually called the Law of Excluded Middle, but which I prefer to call the *Law of Duality*. Whatever the terms A and B may consist of, it is necessarily true, according to this law, that

A is B or not B ;

in symbols

$$A = AB +\!\!\!\mid\, Ab.$$

If any third term C enters into a problem, it is equally certain that

$$A = AC +\!\!\!\mid\, Ac\,;$$

and combining these two developments, we have

$$A = ABC +\!\!\!\mid\, ABc +\!\!\!\mid\, AbC +\!\!\!\mid\, Abc.$$

The same process of subdivision can be carried on *ad infinitum* with respect to any terms that occur; and this Indirect Method of Inference, which I have described in the books mentioned, consists in determining the possible existence of the various alternatives thus produced. The nature and procedure of this method will, as far as possible, be rendered apparent in the mode of treating numerical questions. It has also been partially explained to the Society, in connection with the logical abacus, in which the working of the method is mechanically represented (*Proceedings of the Manch. Lit. and Phil. Soc.*, 3d April 1866, p. 161; see also *Philosophical Transactions*, 1870, p. 497).

10. The data of any problem in numerically definite logic will be of two kinds—

1. The logical conditions governing the combinations of certain qualities or classes of things, expressed in propositions.
2. The numbers of individuals in certain logical classes existing under those conditions.

The *quæsitum* of the problem will be to determine the numbers of individuals in certain other logical classes existing under the same logical conditions, so far as such numbers are rendered determinable by the data. The usefulness of the method will, indeed, often consist in showing whether or not the magnitude of a class is determined or not, or in indicating what further hypotheses or data are required. It will appear, too, that where an exact result

NUMERICALLY DEFINITE REASONING

is not determinable we may yet assign limits within which an unknown quantity must lie.

11. Let us suppose, as an instance, that in a certain statistical investigation, among 100 A's there are found 45 B's and 53 C's; that is to say, in 45 out of 100 cases where A occurs B also occurs, and in 53 cases C occurs. Suppose it to be also known that wherever B is, C also necessarily exists. The data then are as follow—

$$\text{Numerical equations} \begin{cases} (A) = 100 & . \quad . \quad . \quad (1) \\ (B) = 45 & . \quad . \quad . \quad (2) \\ (C) = 53 & . \quad . \quad . \quad (3) \end{cases}$$

Logical equation . $B = BC$.

Let it be required to determine
(1) The number of cases where C exists without B.
(2) The number of cases where neither B nor C exists.

The logical equation asserts that the class B is identical with the class BC, which is the true mode of asserting that all B's are C's. Two distinct results follow from this, namely :—1st, that the number of the class BC is identical with the number of the class B; and 2d, that there are no such things as B's which are not C's.

The logical equation is thus exactly equivalent to two additional numerical equations, namely,

$$(B) = (BC) \quad . \quad . \quad . \quad (4)$$
$$(Bc) = 0 \quad . \quad . \quad . \quad (5)$$

We have now full means of solving the problem; for, by the law of duality,

$$(C) = (BC) + (bC)$$

By (4)
$$= (B) + (bC).$$

Thus
$$53 = 45 + (bC),$$

whence
$$(bC) = 8,$$

which is the first quæsitum.

To obtain the second, the number of Abc's, we have

$$(A) = (ABC) + (ABc) + (AbC) + (Abc)$$
$$100 = 45 + 0 + 8 + (Abc)$$

Hence

$$(Abc) = 47.$$

I now proceed to exemplify the use of the method by applying it to examples drawn chiefly from previous writers.

12. Professor De Morgan suggests the following as an argument which cannot be put into any ordinary form of the syllogism.[1]

'For every man in the house there is a person who is aged; some of the men are not aged. It follows, that some persons in the house are not men.'

This argument proceeds, as I conceive, not by any form of syllogism, but by a pair of simple equations. Taking

$$A = \text{man},$$
$$B = \text{aged person},$$

and putting w, w' for unknown and indefinite numbers, the first premise gives the equation

$$(A) = (B) - w \qquad . \qquad . \qquad . \qquad (1)$$

meaning that the number of aged persons equals or exceeds the number of men. The second statement may be put in this form,

$$(Ab) = w' \qquad . \qquad . \qquad . \qquad (2)$$

which implies that there is a certain indefinite number of men who are not aged.

Develop A and B in (1) by the law of duality, and we have

$$(AB) + (Ab) = (AB) + (aB) - w.$$

Subtract (AB) from both sides, and insert for (Ab) its value in (2), and we have

$$(aB) = w + w' \qquad . \qquad . \qquad . \qquad (3)$$

[1] *Syllabus of a proposed System of Logic*, 1860, p. 29.

which proves that there are some aged persons in the house who are not men, and assigns their quantity, so far as it can be assigned. The number of such persons we learn is at least equal to the number of men who are not aged, and exceeds it by w—that is, the excess of the number of aged persons over the men, if such excess exists, which the premises do not determine.

Adding (ab) to both sides of (3) we get

$$(a) = w + w' + (ab);$$

but this expression contains two unknown quantities, namely, w and (ab). As no quantity can be intrinsically negative, w' is the lowest limit of the number of persons who are not men; and the number is to be increased by w, if it have value, and also by the number of persons, if such there be, who are neither men nor aged.

13. The most celebrated instance to which this method can be applied is one also proposed by Professor De Morgan,[1] and discussed by Boole.[2] It is as follows—

Most B's are A's	. . .	(1)
Most B's are C's	. . .	(2)
Therefore some C's are A's	. . .	(3)

Here, of course, *most* means more than half, and is one of the few quantitative expressions used in ordinary language. We can easily represent the two premises in the form

$$(AB) = \frac{(B)}{2} + w \quad . \quad . \quad . \quad (1)$$

$$(BC) = \frac{(B)}{2} + w' \quad . \quad . \quad . \quad (2)$$

To deduce the conclusion, we must add these equations together, thus,

$$(AB) + (BC) = (B) + w + w'.$$

[1] *Formal Logic*, p. 163.
[2] *Trans. of the Cambridge Philosophical Society*, vol. xi, part ii, p. 1.

Developing the logical terms on each side, we have

$$(ABC)+(ABc)+(ABC)+(aBC) = (ABC)+(ABc)+(aBC) \\ +(aBc)+w+w',$$

Subtracting the common terms, there remains

$$(ABC) = w+w'+(aBc).$$

The meaning of this conclusion is, that there must be some C's which are A's, amounting to at least the sum of the quantities w and w', the unknown excesses beyond half the B's which are A's and C's. The number (aBc) is wholly undetermined by the premises, but it cannot be negative; in proportion as its amount is greater, so is the number of the ABC's. The conclusion, in short, is that $w+w'$ is the lower limit of (ABC).

14. The above problem is only one case of a more general problem, which may be stated as follows:—Given the numbers of three classes of objects, A, B, and C, to determine what circumstances or conditions will necessitate the existence of a class ABC.

This may be solved very simply

$$(B)+(C)-(A) = (ABC)+(ABc)+(ABC)+(AbC)-(A) \\ = (ABC)-(Abc),$$
$$(ABC) = (B)+(C)-(A)+(Abc).$$

It is evident that the number of ABC's is indeterminate, because there is no condition to determine (Abc). But reducing this to its minimum, zero, we learn that the lower limit of (ABC) is the excess of the sum of B's and C's over A's. As no result can be negative, we also learn that if $(Abc) = 0$, then (A) cannot exceed (B)+(C).

15. This method gives us a clear view of the conditions of any logical argument. Take a syllogism in Barbara, thus—

> Every A is B : in symbols $A = AB$.
> Every B is C   ,,   ,, $B = BC$.
> ∴ Every A is C   ,,   ,, $A = AC$.

What additional information do we require in order to determine the number of all the classes of objects concerned?

There are altogether eight conceivable combinations of A, B, C, and their negatives, a, b, c, according to the laws of thought; but of these, four combinations are rendered impossible by the premises, so that we have four quantities assigned—

$$(ABc) = 0, \qquad (Abc) = 0,$$
$$(AbC) = 0, \qquad (aBc) = 0.$$

There remain, then, four unknown quantities; and unless we have these assigned directly or indirectly, we do not really know the relative numbers of the classes. But the numbers of any four existing classes may, by a proper arrangement of equations, be made to yield the number of any other existing class. Thus, if

$$(A) = 93 \qquad (C) = 190$$
$$(aBC) = 5 \qquad (abc) = 4,$$

we may draw the following conclusions—

$$(abC) = (C) - (A) - (aBC) = 190 - 93 - 5 = \mathbf{92};$$
$$(B) = (ABC) + (aBC) = 93 + 5 = 98.$$

16. It is interesting to compare my mode of treating numerically definite propositions with the earlier mode of Professor De Morgan. Taking X, Y, and Z to be the three terms of the syllogism, he adopts[1] the following notation:—

u = whole number of individuals in the universe of the problem.
x = number of X's.
y = number of Y's.
z = number of Z's.

Making m denote any positive number, mXY means that *m or more* X's are Y's. Similarly uYZ means that *u or*

[1] *Syllabus*, p. 27. Mr. De Morgan denotes negative terms by small Roman letters, for which I have substituted italic letters.

more Y's are Z's. Smaller letters denote the negatives of the larger ones, somewhat as in my system. Thus mXy means that m or more X's are not Y's, and so on.

From the two premises

$$m\text{XY and } n\text{YZ},$$

Mr. De Morgan draws the two distinct conclusions

$$(m+n-y)\text{XZ, and } (m+n+u-x-y-z)xy.$$

Let us consider what results are given by my own notation. The premises may be represented by the equations

$$(\text{XY}) = m + m' \qquad (\text{YZ}) = n + n',$$

where m and n are the same quantities as in Mr. De Morgan's system, and m' and n' two unknown but positive quantities, indicating that the number of XY's is m *or more*, and the number of YZ's is n *or more*.

The possible combinations of the three terms X, Y, Z, and their negatives are eight in number, namely—

XYZ,	xYZ,
XYz,	xYz,
XyZ,	xyZ,
Xyz,	xyz,

and these altogether constitute the universe, of which the number is n. The problem is at once seen to be indeterminate in reality; for there are eight classes, of which the number would have to be determined, and there are only six known quantities, namely, u, x, y, z, m, and n, by which to determine them. Accordingly we find that Mr. De Morgan's conclusions, though not absolutely erroneous, have little or no meaning. From the premises he infers that $(m+n-y)$ *or more* X's are Z's. Now

$$m+n-y = (\text{XY}) + (\text{YZ}) - \text{Y}$$
$$= (\text{XYZ}) - (x\text{Y}z).$$

Thus Mr. De Morgan represents the number of the whole

class, XZ, by a quantity indefinitely less than its own part, XYZ. It is quite true that if the second side $(XYZ)-(xYz)$ of this equation has value, there must be at least this number of X's which are Z's; but as (xYz) may exceed (XYZ) in any degree, this may give zero or a negative result, while there is really a large number of XZ's. The true and complete expression for the number of XZ's is found as follows—

$$(XZ) = (XYZ)+(XyZ)$$
$$= (XYZ)+(XYz)+(XYZ)+(xYZ)-(Y)+(XyZ)+(xYz)$$
$$= m+m'+n+n'-y+(XyZ)+(xYz).$$

Among these seven quantities, only m, n, and y are definitely given. The two m' and n' are two indefinite quantities, expressing the uncertainty in the number of XY's and YZ's, while there are two other unknown quantities, the numbers of XyZ's and xYz's arising in the course of the problem.

17. Mr. De Morgan's second conclusion, that the number of not-X's which are not Y's is

$$(m+n+u-x-y-z)$$

or more, may be examined in like manner. By developing the classes numbered in each of these quantities, and striking out the redundant terms, we obtain $(xyz)-(XyZ)$, in which the term (XyZ) is wholly undetermined. Here, again, we have as the lower limit of the class xz a quantity indeterminately less than its own part xyz. The number (xz) may accordingly be of any magnitude, while the limit here assigned to it is zero, or even negative.

Exactly similar remarks may be made concerning the other conclusions which Mr. De Morgan draws. Thus, from mXy and nYz (mX's or more are not Y's, and nY's or more are Z's) he infers

$$(m+n-x)xZ \text{ and } (m+n-z)Xz.$$

But it will be found by analysis that the first of these results has the following meaning :—

$$(xZ)\overset{=}{>}(x\mathrm{Y}Z) - (\mathrm{XY}z);$$

that is to say, the lower limit of the class xZ is a part of itself, $x\mathrm{Y}Z$, diminished by the number of another class $\mathrm{XY}z$.

While believing, however, that Mr. De Morgan's mode of treating the subject admits of improvement, it is impossible that I should undervalue the extraordinary acuteness and originality of his writings on this and many other parts of formal logic. Time is required to reveal the wealth of thought which he has embodied in his *Formal Logic*, and in his *Logical Memoirs* published by the Cambridge Philosophical Society.

18. In Mr. De Morgan's third paper on the syllogism[1] he puts the syllogism in the following form:—'If the fractions a and β of the Y's be severally A's and B's, and if $a+\beta$ be greater than unity, it follows that some A's are B's. . . . The logician demands $a=1$ or $\beta=1$, or both; he can then infer.' These arguments are readily represented in my notation as follows—

The premises are $\quad a \cdot (\mathrm{Y}) = (\mathrm{AY}),$
$\qquad\qquad\qquad\qquad \beta \cdot (\mathrm{Y}) = (\mathrm{BY}).$

Hence

$$(a+\beta)(\mathrm{Y}) = (\mathrm{AY}) + (\mathrm{BY})$$
$$= (\mathrm{ABY}) + (\mathrm{A}b\mathrm{Y}) + (\mathrm{ABY}) + (a\mathrm{BY}),$$
$$(a+\beta)(\mathrm{Y}) - (\mathrm{Y}) = (\mathrm{ABY}) - (ab\mathrm{Y}),$$

or

$$(\mathrm{ABY}) = (a+\beta - 1)(\mathrm{Y}) + (ab\mathrm{Y}).$$

From this we learn that the number of A's which are B's, *because they are Y's*, is the fraction $(a+\beta-1)$ of the Y's together with the undetermined number $(ab\mathrm{Y})$, which cannot be negative. Hence if $a+\beta > 1$, the second side has a positive value, and there must be some A's which are B's. If $a=1$, then this number is $\beta \cdot (\mathrm{Y})$, or if $\beta=1$, it is $a \cdot (\mathrm{Y})$, since $(ab\mathrm{Y})$ then $= 0$. If $a=1$ and $\beta=1$, then obviously $(\mathrm{ABY}) = (\mathrm{Y})$.

[1] *Cambridge Phil. Trans.* vol. x, part i, p. 8.

19. In Mr. Mill's chapter 'On Chance and its Eliminations,'[1] occurs a problem concerning the coexistence of two phenomena, in which he asserts the general proposition 'that, if A occurs in a larger proportion of the cases where B is than of the cases where B is not, then will B also occur in a larger proportion of the cases where A is than of the cases where A is not.'

This proposition is not proved by Mr. Mill, nor do I remember seeing any proof of it; and it is not, to my mind, self-evident. The following, however, is a proof of its truth, and is the shortest proof I have been able to find.

The condition of the problem may be expressed in the inequality

$$\frac{(AB)}{(B)} > \frac{(Ab)}{(b)},$$

or reciprocally in the inequality

$$\frac{(B)}{(AB)} < \frac{(b)}{(Ab)}.$$

Subtracting unity from each side, and simplifying, we have

$$\frac{(aB)}{(AB)} < \frac{(ab)}{(Ab)}.$$

Multiplying each side of this inequality by $\frac{(Ab)}{(aB)}$ we obtain

$$\frac{(Ab)}{(AB)} < \frac{(ab)}{(aB)}.$$

Restoring unity to each side, and simplifying

$$\frac{(A)}{(AB)} < \frac{(a)}{(aB)},$$

or reciprocally

$$\frac{(AB)}{(A)} > \frac{(aB)}{(a)},$$

---

[1] *System of Logic*, fifth ed. vol. ii, p. 54.

which expresses the result to be proved, namely, that B occurs in a larger proportion of the cases where A is than of the cases where A is not.

20. The examples hitherto considered have been mostly free from logical conditions; that is to say, the classes of objects have been supposed capable of combination or coincidence in all conceivable ways. We will briefly examine the effects of certain simple logical conditions.

If there be two terms A and B, and one condition, *all A's are B's*, symbolically expressed in the equation

$$A = AB,$$

then there will be three possible classes to be determined, namely,

$$AB,$$
$$aB,$$
$$ab,$$

and we shall require three assigned quantities. If we have (U) = whole number of objects, with (A) and (B), then

$$(AB) = (A)$$
$$(aB) = (B) - (A)$$
$$(ab) = (U) - (B).$$

21. If with two terms, A and B, the logical condition be $A = B$, there will remain two classes only, AB and ab, and two assigned quantities only will be required. The same would happen with any of the conditions $A = b$, $a = B$, or $a = b$.

22. In any problem involving three terms or classes of things, say A, B, and C, there arise eight conceivable classes, the numbers of which may have to be determined. Various logical conditions, however, greatly reduce the numbers. Thus the two conditions

$$A = B = C$$

leave only two possible classes, ABC, and *abc*.

The two conditions

$$A = AB \text{ and } B = BC$$

leave four classes,

ABC, aBC, abC, and abc.

23. The two conditions A is B or C, but B cannot be C, symbolically expressed

$$A = AB \cdot |\cdot AC, \qquad\qquad B = B c,$$

leave five classes,

ABc, AbC, aBc, abC, abc.

24. These few examples illustrate the way in which the indirect method of inference, described in my *Pure Logic*, determines the number of possible classes which may exist under certain logical conditions, and thus enables us to ascertain at once whether there are data sufficient to determine their magnitude. Various examples of the process may be found in the work referred to.

25. My formulæ will also, I believe, be found to yield all the aid to the calculation of probabilities which can be expected from the science of logic. When the combinations of events are not governed by any special logical conditions, the application of the logical formulæ to probabilities is exceedingly simple. It is only necessary in the logical formula to substitute for each term its probability of occurrence, and to multiply or add as the logical signs indicate.

Thus, if p is the probability of the event A happening, and q of B, then pq is the probability of the conjunction of events AB happening; similarly the probability of A not happening, that is, of a happening, is $1-p$; of b, $1-q$. According we have the following:—

Probability of AB $= pq$.
  ,,   ,,   Ab $= p(1-q)$.
  ,,   ,, aB $= (1-p)q$.
  ,,   ,, ab $= (1-p)(1-q)$.

26. In Chapter XVIII of his *Laws of Thought*, Boole has given several examples of the application of his very complicated General Method of Probabilities. Of these examples my notation will give a vastly simpler solution, as I proceed to show.

Boole's third example is as follows (p. 279)—

'The probability that a witness, A, speaks the truth is p, the probability that another witness, B, speaks the truth is q, and the probability that they disagree in a statement is r. What is the probability that if they agree in a statement, their statement is true?'

This is solved in the simplest possible manner. Let

a = prob. of A and B both speaking truth.
β = prob. of A but not B ,, ,,
γ = prob. of not A but B ,, ,,
δ = prob. of neither A nor B ,, ,,

Then we have the following data:—

Prob. of A speaking truth $= a + \beta = p$.
,, B ,, ,, $= a + \gamma = q$.
Prob. that they disagree $= \beta + \gamma = r$.

As it is certain that one or other of the alternatives must happen, we have the condition

$$a + \beta + \gamma + \delta = 1.$$

These four equations are sufficient to determine all the four unknown quantities by ordinary algebra. Thus

$$a = \frac{p+q-r}{2},$$
$$\delta = 1 - (a + \beta + \gamma) = 1 - \frac{p+q-r}{2} - r,$$
$$= 1 - \frac{p+q+r}{2}.$$

Now, the probability required is, that if A and B agree in

a statement their statement is true. By the principles of probability this is $\dfrac{a}{a+\delta}$; and inserting the above values of a and δ we have

$$\frac{a}{a+\delta}=\frac{p+q-r}{2(1-r)},$$

which is the same as the result which Boole obtained in a much more complicated manner. This verifies the anticipations both of Boole himself (p. 281) and of Mr. Wilbraham,[1] in his criticism on Boole's *Method of Probabilities,* 'that the really determinate problems solved in the book, as 2 and 3 of Chapter XVIII, might be more shortly solved.' Boole remarks, indeed, that they do not fall directly within the scope of known methods; but I conceive that my logical symbols and method furnish all that is required.

27. In a similar manner we may solve the second of Boole's examples referred to by Mr. Wilbraham; this is as follows—

'The probability that one or both of two events happen is p, that one or both of them fail is q. What is the probability that only one of these happens?'

Using a, β, γ, δ to denote the probabilities of the four obvious conjunctions of events, as before, we have the data,

$$a+\beta+\gamma=p,$$
$$\beta+\gamma+\delta=q,$$
$$a+\beta+\gamma+\delta=1.$$

The probability required is $\beta+\gamma$, and

$$\beta+\gamma = q-\delta = q-1+a+\beta+\gamma$$
$$= q-1+p.$$

This is Mr. Boole's result, obtained by him in a much more complicated manner.

28. This simple substitution of the probability of an event for its logical symbol cannot be valid, however, if

---

[1] *Philosophical Magazine,* 4th series, vol. vii, p. 465; vol. viii, p. 91.

there be any connection between the events which renders one more or less likely to happen when the other happens. The probabilities of A and B being p and q, the probability of AB is pq, under the supposition that B is just as likely to happen when A happens as when A does not happen and similarly that A is just as likely to happen when B does as when B does not; in short, that they are *independent events*. As a case where we are not to assume logical independence, we may take the following example from Boole's work (p. 276):—

Example 1. 'The probability that it thunders upon a given day is p, the probability that it both thunders and hails is q; but of the connection of the two phenomena of thunder and hail nothing further is supposed to be known. Required the probability that it hails on the proposed day.'

Let A mean that it thunders
     B   „   „   hails;

Then there are four possible events, AB, Ab, aB, ab.
The probabilities given are—

$$\text{Prob. of A} = p,$$
$$\text{,, \quad AB} = q.$$

The probability required is that of B, which is evidently

$$\text{Prob. of AB} + \text{prob. of } a\text{B}.$$

Now the probability of AB is given, but the probability of aB is not given, and we cannot assume it to be $(1-p) \times$ (prob. of B), because we are told that nothing is known of the connection of the phenomena, which implies that they may have some connection by causation, so that the non-occurrence of A will alter the probability of the occurrence of B. The prob. of aB is therefore unknown, except that it is the prob. of a multiplied by the unknown prob. that, if a occurs, B occurs with it, as Boole points out. Hence the only possible answer is the same as Boole's

$$\text{Prob. of B} = q + (1-p)c,$$

c being an unknown quantity, of course not exceeding unity. Making c successively $= 1$ and 0, the major and minor limits of the probability are evidently $q + 1 - p$ and q.

Were the events A and B independent, we should have

Prob. of B $= q + (1 - p)$ (prob. of B),

$$= \frac{q}{1 - (1 - p)} = \frac{q}{p}.$$

29. It may be truly remarked of what is given in this paper, that all the results can be reached by the exertion of common sense, or by ordinary mathematical calculation; and I do not doubt that problems combining logical and mathematical conditions of a more complicated character have been solved, especially in the *Theory of Probabilities*, by those who were unconscious of using any peculiar logical method; but what I claim for my logical method and notation is, that it is in no sense or way peculiar, but represents truthfully and completely the natural course of intelligent thought. The indirect method, first explained in 1864 in my *Pure Logic*, embodied in the mechanical device called the Logical Abacus, explained to the Society in April 1866, and further exemplified in the Logical Machine lately brought before the Royal Society, represents the exhaustive and necessary classification of objects which the mind must make under any logical conditions. Of previous systems, Boole's mathematical method could alone be said to do this; and his method was deformed by needless obscurity, and by at least one deep-seated error. It has been my purpose in this paper to exemplify the way in which a true and simple logical method lends its aid to all such mathematical problems as involve logical considerations. The number of such problems requiring solution is not great, unless, perhaps, in the theory of probabilities; but I believe that in the progress of science the number will probably increase. And whether this be so or not, we must not estimate the value of a theory by its immediate practical results.

30. Logical method must undoubtedly be the root of all

scientific demonstration, and of all sound thought in the common affairs of life; yet we find the most opposite and contradictory opinions held by different logicians as to the nature of the reasoning process. Metaphysical speculation will never remedy the present deplorable condition of the science; for it is metaphysical speculation which has mystified the subject, and rendered it the laughing-stock of scientific men. Antiquarian research into the errors of earlier logicians, in which some logicians still exclusively employ themselves, will only add to the perplexity and obscurity. I hold that logic can only be regenerated by those who will render themselves acquainted with the exact methods of research which lead to undoubted truths in the mathematical and physical sciences. Logic, in short, must be dissociated from metaphysics, with which it has no necessary connection, and must become an exact science. We must therefore seek in every way to connect it with the other exact sciences. In this paper I have attempted to show that questions do exist in which logical and numerical methods coalesce and lend mutual aid.

PART II
JOHN STUART MILL'S PHILOSOPHY TESTED

PORTIONS OF AN EXAMINATION OF
JOHN STUART MILL'S PHILOSOPHY

I

ON GEOMETRICAL REASONING

DURING the last few weeks the correspondence columns of the *Spectator* have contained letters on the subject of the late Mr. Mill's opinions about the Immortality of the Soul. The discussion began with a letter, in which an anonymous writer, G. S. B., asserted that Mill spoke of immortality as *probably an illusion*, although morally so valuable an illusion that it is better to retain it. He went on to say, 'It is surely time that all this scientific shuffling and intellectual dishonesty—for it is nothing else—should be exposed and exploded.'

An ardent admirer of Mill was not unnaturally stung by this remark, and replied in a letter, ably and warmly vindicating Mill's truthfulness and 'scrupulous accurateness.' After showing, as he thinks, that Mill never tried to uphold any illusion, he thus concludes—

'It is very difficult to misunderstand Mr. Mill, so anxious was he always to be clear, to be just, to keep back nothing, to examine both sides, to overstate nothing and to understate nothing, so sensitively honourable was his mind, so transparently honest his style. But these are commonplaces with respect to him. I am content to contrast the scrupulous accurateness of Mr. Mill with what appears of that quality in "G. S. B."'

In the *Spectator* of the following week (27th October), I took the opportunity to express my dissent from both the correspondents, saying—

'I do not like the expression "scientific shuffling and intellectual dishonesty" which G. S. B. has used, for fear it should imply that Mill knowingly misled his readers. It is impossible to doubt that Mill's mind was "sensitively honourable," and, whatever may be his errors of judgment, we cannot call in question the perfect good faith and loftiness of his intentions. On the other hand, it is equally difficult to accept what Mr. Malleson says as to the "scrupulous accurateness" of Mill's *Essays on Religion*. He was scrupulous, but the term "accurateness," if it means "logical accurateness," cannot be applied to his works by any one who has subjected them to minute logical criticism.'

I then pointed out that, in pp. 109 and 103 of his *Essays on Religion*, Mill gives two different definitions or descriptions of religion. In the first he says that

'the essence of religion is the strong and earnest direction of the emotions and desires towards an ideal object, recognised as of the highest excellence, and as rightfully paramount over all selfish objects of desire.'

In the second statement he says—

'Religion, as distinguished from poetry, is the product of the craving to know whether these imaginative conceptions have realities answering to them in some other world than ours.'

A week afterwards Mr. Malleson made an ingenious attempt to explain away or to palliate the obvious discrepancy by reference to the context. I do not think that any context can remove the discrepancy; in the one case the object of desire is an *ideal* object; in the other case the *craving*, which I presume means a strong desire, is towards realities in some other world; and the difference between ideal and real is too wide for any context to bridge over. Besides, I will ultimately give reasons for holding that Mill's text cannot be safely interpreted by the context, because there is no certainty that in his writings the same line of thought is steadily maintained for two sentences in succession.

Mill's *Essays on Religion* have been the source of perplexity to numberless readers. His greatest admirers have been compelled to admit that in these essays even Mill seems now and then to play with a word, or unconsciously to mix up two views of the same subject. It has been urged, indeed, by many apologists, including Miss Helen Taylor, their editor, that Mill wrote these essays at wide intervals of time, and was deprived, by death, of the opportunity of giving them his usual careful revision. This absence of revision, however, applies mainly to the third essay, while the discrepant definitions of religion were quoted from the second essay. Moreover, lapse of time will not account for inconsistency occurring between pages 103 and 109 of the same essay. The fact simply is that these essays, owing to the exciting nature of their subjects, have received a far more searching and hostile criticism than any of his other writings. Thus inherent defects in his intellectual character, which it was a matter of great difficulty to expose in so large a work as the *System of Logic*, were readily detected in these brief, candid, but most ill-judged essays.

But, for my part, I will no longer consent to live silently under the incubus of bad logic and bad philosophy which Mill's Works have laid upon us. On almost every subject of social importance — religion, morals, political philosophy, political economy, metaphysics, logic—he has expressed unhesitating opinions, and his sayings are quoted by his admirers as if they were the oracles of a perfectly wise and logical mind. Nobody questions, or at least ought to question, the force of Mill's style, the persuasive power of his words, the candour of his discussions, and the perfect goodness of his motives. If to all his other great qualities had been happily added logical accurateness, his writings would indeed have been a source of light for generations to come. But in one way or another Mill's intellect was wrecked. The cause of injury may have been the ruthless training which his father imposed upon him in tender years; it may have been Mill's own lifelong attempt to reconcile a false empirical philosophy with conflicting truth. But however it arose, Mill's mind was essentially illogical.

Such, indeed, is the intricate sophistry of Mill's principal writings, that it is a work of much mental effort to trace out the course of his fallacies. For about twenty years past I have been a more or less constant student of his books: during the last fourteen years I have been compelled, by the traditional requirements of the University of London, to make those works at least partially my text-books in lecturing. Some ten years of study passed before I began to detect their fundamental unsoundness. During the last ten years the conviction has gradually grown upon my mind that Mill's authority is doing immense injury to the cause of philosophy and good intellectual training in England. Nothing surely can do so much intellectual harm as a body of thoroughly illogical writings, which are forced upon students and teachers by the weight of Mill's reputation, and the hold which his school has obtained upon the universities. If, as I am certain, Mill's philosophy is sophistical and false, it must be an indispensable service to truth to show that it is so. This weighty task I at length feel bound to undertake.

The mode of criticism to be adopted is one which has not been sufficiently used by any of his previous critics. Many able writers have defended what they thought the truth against Mill's errors; but they confined themselves for the most part to skirmishing round the outworks of the Associationist Philosophy, firing in every here and there a well-aimed shot. But their shots have sunk harmlessly into the sand of his foundations. In order to have a fair chance of success, different tactics must be adopted; the assault must be made directly against the citadel of his logical reputation. His magazines must be reached and exploded; he must be hoist, like the engineer, with his own petard. Thus only can the disconnected and worthless character of his philosophy be exposed.

I undertake to show that there is hardly one of his more important and peculiar doctrines which he has not himself amply refuted. It will be shown that in many cases it is impossible to state what his doctrine is, because he mixes up two or three, and, in one extreme case, as many as six different and inconsistent opinions. In several important cases, the view which he professes to uphold is the direct opposite of what he really upholds. Thus, he clearly reprobates the doctrine of Free Will, and expressly places himself in the camp of Necessity; but he objects to the name Necessity, and explains it away so ingeniously, that he unintentionally converts it into Free Will. Again, there is no doubt that Mill wished and believed himself to be a bulwark of the Utilitarian Morality; he prided himself on the invention, or at least the promulgation, of the name Utilitarianism; but he expounded the doctrines of the school with such admirable candour, that he converted them unconsciously into anything rather than the doctrines of Paley and Bentham.

As regards logic, the case is much worse. He affected to get rid of universal reasoning, which, if accomplished, would be to get rid of science and logic altogether; of course he employed or implied the use of universals in almost every sentence of his treatise. He overthrew the syllogism on the ground of *petitio principii*, and then immediately set it up again as an indispensable test of good reasoning. He defined logic as the Science of Proof, and then recommended a loose kind of inference from particulars to particulars, which he allowed was not conclusive, that is, could prove nothing. Though inconclusive, this loose kind of inference was really the basis of conclusive reasoning. Then, again, he founded induction upon the law of causation, and at the same time it was his express doctrine that the law of causation was learned by induction. What he meant exactly by this law of causation it is impossible to say. He affirms and denies the plurality of causes. Sometimes the sequence of causation is absolutely invariable, sometimes it is conditional. Generally, the law of causation is spoken of as Universal, or as universal throughout nature; yet in one passage (at the end of Book III, chap. xxi) he makes a careful statement to the opposite effect, and this statement, subversive as it is of his whole system of induction, has appeared in all editions from the first to the last. On such fundamental questions as the meaning of propositions, the nature of a class, the theory of probability, etc., he is in error where he is not in direct conflict with himself. But the indictment is long enough already; there is not space in this article to complete it in detail. To sum up, there is nothing in logic which he has not touched, and he has touched nothing without confounding it.

To establish charges of this all-comprehensive character will of course require a large body of proof. It will not be sufficient to take a few of Mill's statements and show that they are mistaken or self-inconsistent. Any writer may now and then fall into oversights, and it would be manifestly unfair to pick a few unfortunate passages out of a work of considerable extent, and then hold them up as specimens of the whole. On the other hand, in order to overthrow a philosopher's system, it is not requisite to prove his every statement false. If this were so, one large treatise would require ten large ones to refute it. What is necessary is to select a certain number of his more prominent and peculiar doctrines, and to show that, in their treatment, he is illogical. In this article, I am, of course, limited in space and can apply only one test, and the subject which I select for treatment is Mill's doctrines concerning geometrical reasoning.

The science of geometry is specially suited to form a test of the empirical philosophy. Mill certainly regarded it as a crucial instance, and devoted a considerable part of his *System of Logic* to proving that geometry is a *strictly physical science*, and can be learnt by direct observation and induction. The particular nature of his doctrine, or rather *doctrines*, on this subject will be gathered as we proceed. Of course, in this inquiry I must not abstain from a searching or even a tedious analysis, when it is requisite for the due investigation of Mill's logical method; but it will rarely be found necessary to go beyond elementary mathematical knowledge which almost all readers of the *Contemporary Review* will possess.

As a first test of Mill's philosophy I propose this simple question of fact—Are there in the material universe such things as perfectly straight lines? We shall find that Mill returns to this question a categorical negative answer. There exist no such things as perfectly straight lines. How then can geometry exist, if the things about which it is conversant do not exist? Mill's ingenuity seldom fails him. Geometry, in his opinion, treats not of things as they are in reality, but as we suppose them to be. Though straight lines do not exist, we can experiment in our minds upon straight lines, as if they did exist. It is a peculiarity of geometrical science, he thinks, thus to allow of *mental experimentation*. Moreover, these mental experiments are just as good as real experiments, because we know that the imaginary lines exactly resemble real ones, and that we can conclude from them to real ones with quite as much certainty as we conclude from one real line to another. If such be Mill's doctrines, we are brought into the following position :—

1. Perfectly straight lines do not really exist.

2. We experiment in our minds upon imaginary straight lines.

3. These imaginary straight lines exactly resemble the real ones.

4. If these imaginary straight lines are not perfectly straight, they will not enable us to prove the truths of geometry.

5. If they are perfectly straight, then the real ones, which *exactly* resemble them, must be perfectly straight: *ergo*, perfectly straight lines do exist.

It would not be right to attribute such reasoning to Mill without fully substantiating the statements. I must therefore ask the reader to bear with me while I give somewhat full extracts from the fifth chapter of the second book of the *System of Logic*.

Previous to the publication of this 'system,' it had been generally thought that the certainty of geometrical and other mathematical truths was a property not exclusively confined to these truths, but nevertheless existent. Mill, however, at the commencement of the chapter, altogether calls in question this supposed certainty, and describes it as an *illusion,* in order to sustain which it is necessary to suppose that those truths relate to, and express the properties of, purely imaginary objects. He proceeds [1]—

'It is acknowledged that the conclusions of geometry are deduced, partly at least, from the so-called Definitions, and that those definitions are assumed to be correct descriptions, as far as they go, of the objects with which geometry is conversant. Now we have pointed out that, from a definition as such, no proposition, unless it be one concerning the meaning of a word, can ever follow, and that what apparently follows from a definition, follows in reality from an implied assumption that there exists a real thing conformable thereto. This assumption, in the case of the definitions of geometry, is false:[2] there exist no real things exactly conformable to the definitions. There exist no points without magnitude; no lines without breadth, nor perfectly straight; no circles with all their radii exactly equal, nor squares with all their angles perfectly right. It will perhaps be said that the assumption does not extend to the actual, but only to the possible, existence of such things. I answer that, according to any test we have of possibility, they are not even possible. Their existence, so far as we can form any judgment, would seem to be inconsistent with the physical constitution of our planet at least, if not of the universe.'

About the meaning of this statement no doubt can arise. In the clearest possible language Mill denies the existence of perfectly straight lines, so far as any judgment can be formed, and this denial extends, not only to the actual, but the possible, existence of such lines. He thinks that they

[1] Book ii, chap. v, sec. 1, near the commencement of the second paragraph.
[2] The word *false* occurs in the editions up to at least the fifth edition. In the latest or ninth edition I find the words, *not strictly true,* substituted for false.

seem to be inconsistent with the physical constitution of our planet, if not of the universe. Under these circumstances, there naturally arises the question, What does geometry treat? A science, as Mill goes on to remark, cannot be conversant with nonentities; and as perfectly straight lines and perfect circles, squares, and other figures do not exist, geometry must treat such lines, angles, and figures as do exist, these apparently being imperfect ones. The definitions of such objects given by Euclid, and adopted by later geometers, must be regarded as some of our first and most obvious generalisations concerning those natural objects. But then, as the lines are never perfectly straight nor parallel, in reality, the circles not perfectly round, and so on, the truths deduced in geometry cannot accurately apply to such existing things. Thus we arrive at the necessary conclusion that the peculiar accuracy attributed to geometrical truths is *an illusion*. Mill himself clearly expresses this result [1]—

'The peculiar accuracy, supposed to be characteristic of the first principles of geometry, thus appears to be fictitious. The assertions on which the reasonings of the science are founded, do not, any more than in other sciences, exactly correspond with the fact; but we *suppose* that they do so, for the sake of tracing the consequences which follow from the supposition.'

So far Mill's statements are consistent enough. He gives no evidence to support his confident assertion that perfectly straight lines do not exist; but with the actual truth of his opinion I am not concerned. All that would be requisite to the logician, as such, is that, having once adopted the opinion, he should adhere to it, and admit nothing which leads to an opposite conclusion.

The question now arises in what way we obtain our knowledge of the truths of geometry, especially those very general truths called axioms. Mill has no doubt whatever about the answer. He says [2]—

[1] Book ii, chap. v, sec. 1, at the beginning of the fourth paragraph.
[2] Same chapter, at the beginning of section 4.

'It remains to inquire, What is the ground of our belief in axioms—what is the evidence on which they rest? I answer, they are experimental truths; generalisations from observation. The proposition, Two straight lines cannot enclose a space—or in other words, Two straight lines which have once met, do not meet again, but continue to diverge—is an induction from the evidence of our senses.'

This opinion, as Mill goes on to remark, runs counter to a scientific prejudice of long standing and great force, and there is probably no proposition enunciated in the whole treatise for which a more unfavourable reception was to be expected. I think that the "scientific prejudice" still prevails, but I am perfectly willing to agree with Mill's demand that the opinion is entitled to be judged, not by its novelty, but by the strength of the arguments which are adduced in support of it. These arguments are the subject of our inquiry. Mill proceeds to point out that the properties of parallel or intersecting straight lines are apparent to us in almost every instant of our lives. 'We cannot look at any two straight lines which intersect one another, without seeing that from that point they continue to diverge more and more.'[1] Even Whewell, the chief opponent of Mill's views, allowed that observation *suggests* the properties of geometrical figures; but Mill is not satisfied with this, and proceeds to controvert the arguments by which Whewell and others have attempted to show that experience cannot *prove* the axiom.

The chief difficulty is this: before we can assure ourselves that two straight lines do not enclose space, we must follow them to infinity. Mill faces the difficulty with boldness and candour—

'What says the axiom? That two straight lines *cannot* enclose a space; that after having once intersected, if they are prolonged to infinity they do not meet, but continue to diverge from one another. How can this, in any single case, be proved by actual observation? We may follow the lines to any distance we please;

---

[1] Same section, near the beginning of fourth paragraph.

but we cannot follow them to infinity : for aught our senses can testify, they may, immediately beyond the farthest point to which we have traced them, begin to approach, and at last meet. Unless, therefore, we had some other proof of the impossibility than observation affords us, we should have no ground for believing the axiom at all.

'To these arguments, which I trust I cannot be accused of understating, a satisfactory answer will, I conceive, be found, if we advert to one of the characteristic properties of geometrical forms—their capacity of being painted in the imagination with a distinctness equal to reality : in other words, the exact resemblance of our ideas of form to the sensations which suggest them. This, in the first place, enables us to make (at least with a little practice) mental pictures of all possible combinations of lines and angles, which resemble the realities quite as well as any which we could make on paper ; and in the next place, make those pictures just as fit subjects of geometrical experimentation as the realities themselves ; inasmuch as pictures, if sufficiently accurate, exhibit of course all the properties which would be manifested by the realities at one given instant, and on simple inspection : and in geometry we are concerned only with such properties, and not with that which pictures could not exhibit, the mutual action of bodies one upon another. The foundations of geometry would therefore be laid in direct experience, even if the experiments (which in this case consist merely in attentive contemplation) were practised solely upon what we call our ideas, that is, upon the diagrams in our minds, and not upon outward objects. For in all systems of experimentation we take some objects to serve as representatives of all which resemble them ; and in the present case the conditions which qualify a real object to be the representative of its class, are completely fulfilled by an object existing only in our fancy. Without denying, therefore, the possibility of satisfying ourselves that two straight lines cannot enclose a space, by merely thinking of straight lines without actually looking at them ; I contend, that we do not believe this truth on the ground of the imaginary intuition simply, but because we know that the imaginary lines exactly resemble real ones, and that we may conclude from them to real ones with quite as much certainty as we could conclude from one real line to another. The conclusion, therefore, is still an induction from observation.'[1]

[1] Book ii, chap. v, sec. 5. The passage occurs in the second and third paragraphs.

I have been obliged to give this long extract in full, because, unless the reader has it all freshly before him, he will scarcely accept my analysis. In the first place, what are we to make of Mill's previous statement that the axioms are *inductions from the evidence of our senses?* Mill admits that, for aught our senses can testify, two straight lines, although they have once met, may again approach and intersect beyond the range of our vision. 'Unless, therefore, we had some other proof of the impossibility than observation affords us, we should have no ground for believing the axiom at all.'[1] Probably it would not occur to most readers to inquire whether such a statement is consistent with that made two or three pages before, but on examination we find it entirely inconsistent. Before, the axioms were inductions from *the evidence of our senses;* now, we must have 'some other proof of the impossibility than *observation* affords us.'

This further proof, it appears, consists in the attentive contemplation of mental pictures of straight lines and other geometrical figures. Such pictures, if sufficiently accurate, exhibit, of course, all the properties of the real objects, and in the present case the conditions which qualify a real object to be the representative of its class are completely fulfilled. Such pictures, Mill admits, must be *sufficiently accurate;* but what, in geometry, is sufficient accuracy? The expression is, to my mind, a new and puzzling one. Imagine, since Mill allows us to do so, two parallel straight lines. What is the sufficient accuracy with which we must frame our mental pictures of such lines, in order that they shall not meet? If one of the lines, instead of being really straight, is a portion of a circle having a radius of a hundred miles, then the divergence from perfect straightness within the length of one foot would be of an order of magnitude altogether imperceptible to our senses. Can we, then, detect in the mental picture that which cannot be detected in the sensible object? This can hardly be held by Mill,

---

[1] End of the second paragraph.

because he says, further on, that we are only warranted in substituting observation of the image in our mind for observation of the reality by long-continued experience that the properties of the reality are faithfully represented in the image.

Now, since we may (at least with a little practice) form mental pictures of all possible combinations of lines and angles, we may, I presume, form a picture of lines which are so nearly parallel that they will only meet at a distance of 100,000 miles. If we cannot do so, how can we detect the difference between such lines and those that are actually parallel? Mill meets this difficulty. If two lines meet at a great distance,

'we can transport ourselves thither in imagination, and can frame a mental image of the appearance which one or both of the lines must present at that point, which we may rely on as being precisely similar to the reality. Now, whether we fix our contemplation upon this imaginary picture, or call to mind the generalisations we have had occasion to make from former ocular observation, we learn by the evidence of experience, that a line which, after diverging from another straight line, begins to approach to it, produces the impression on our senses which we describe by the expression, "a bent line," not by the expression, "a straight line." '[1]

In this passage we have somewhat unexpectedly got back to *the senses*. We may call to mind the generalisations from former ocular observation, and we have the evidence of experience to distinguish between the impressions made on our senses by a bent line and a straight line. But what will happen if the bent line be a circle with a radius of a million miles? Have we the evidence of experience that two such lines, which seem to be parallel for the first hundred miles, afterwards begin to approach, and finally intersect. If so, our senses must enable us to see clearly and to exactly measure quantities a hundred miles away. Or again, if there be two lines which close in front of me are

[1] Book ii, chap. v, sec. 5, end of fourth paragraph.

one foot apart, but which a hundred miles away are one foot *plus* the thousandth of an inch apart, they are not parallel. Will my senses enable me to perceive the magnitude of the thousandth part of an inch placed a hundred miles off?

But we have had enough of this trifling. Any one who has the least knowledge of geometry must know that a straight line means a *perfectly* straight line; the slightest curvature renders it not straight. Parallel straight lines mean *perfectly* parallel straight lines; if they be in the least degree not parallel, they will of course meet sooner or later, provided that they be in the same plane. Now Mill said that we get an impression on our senses of a straight line; it is through this impression that we are enabled to form images of straight lines in the mind. We are told,[1] moreover, that *the imaginary lines exactly resemble real ones,* and that it is long-continued observation which teaches us this. It follows most plainly, then, that the impressions on our senses must have been derived from really straight lines. Mill's philosophy is essentially and directly empirical; he holds that we learn the principles of geometry by direct ocular perception, either of lines in nature, or their images in the mind. Now if our observations had been confined to lines which are not parallel, we could by no possibility have perceived, directly and ocularly, the character of lines which are parallel. It follows, that *we must have perceived perfectly parallel lines and perfectly straight lines, although Mill previously told us that he considered the existence of such things to be 'inconsistent with the physical constitution of our planet, at least, if not of the universe.'*

Perhaps it may be replied that Mill simply made a mistake in saying that no really straight lines exist, and, correcting this blunder of fact, the logical contradiction vanishes. Certainly he gives no proper reason for his confident denial of their existence. But merely to strike out a page of Mill's Logic will not vindicate his logical character.

[1] Same section, about thirteen lines from the end of the third paragraph.

How came he to put a statement there which is in absolute conflict with the rest of his arguments? No interval of time, no want of revision, can excuse this inconsistency, for the passage occurs in the first edition of the *System of Logic* (vol. i, p. 297), and reappears unchanged (except as regards one word) in the last and ninth edition. The curious substitution of the words 'not strictly true' for the word 'false' shows that Mill's attention had been directed to the paragraph; and a good many remarks might be made upon this little change of words, were there not other matters claiming prior attention.

We have seen that Mill considers our knowledge of geometry to be founded to a great extent on *mental experimentation*. I am not aware that any philosopher ever previously asserted, with the same distinctness and consciousness of his meaning, that the observation of our own ideas might be substituted for the observation of things. Philosophers have frequently spoken of their ideas or notions, but it was usually a mere form of speech, and their ideas meant their direct knowledge of things. Certainly this was the case with Locke, who was always talking about ideas. Descartes, no doubt, held that whatever we can clearly perceive is true; but he probably meant that it would be logically possible. I do not think that Descartes in his geometry ever got to *mental experimentation*. But however this may be, Mill, of all men, ought not to have recommended such a questionable scientific process, if we may judge from his statements in other parts of the *System of Logic*. The fact is that Mill, before coming to the subject of Geometry, had denounced *the handling of ideas instead of things* as one of the most fatal errors—indeed, as *the cardinal error of logical philosophy*. In the chapter upon the Nature and Import of Propositions,[1] he says—

'The notion that what is of primary importance to the logician in a proposition, is the relation between the two *ideas*

---

[1] Book i, chap. v, sec. 1, fifth paragraph.

corresponding to the subject and predicate (instead of the relation between the two *phenomena* which they respectively express), seems to me one of the most fatal errors ever introduced into the philosophy of Logic; and the principal cause why the theory of the science has made such inconsiderable progress during the last two centuries. The treatises on Logic, and on the branches of Mental Philosophy connected with Logic, which have been produced since the intrusion of this cardinal error, though sometimes written by men of extraordinary abilities and attainments, almost always tacitly imply a theory that the investigation of truth consists in contemplating and handling our ideas, or conceptions of things, instead of the things themselves: a doctrine tantamount to the assertion, that the only mode of acquiring knowledge of nature is to study it at second hand, as represented in our own minds.'

Mill here denounces the *cardinal error* of investigating nature at second hand, as represented in our own minds. Yet his words exactly describe that process of mental experimentation which he has unquestionably advocated in geometry, the most perfect and certain of the sciences.

It may be urged, indeed, with some show of reason, that the method which might be erroneous in one science might be correct in another. The mathematical sciences are called the exact sciences, and they may be of peculiar character. But, in the first place, Mill's denunciation of the handling of ideas is not limited by any exceptions; it is applied in the most general way, and arises upon the general question of the Import of Propositions. It is, therefore, in distinct conflict with Mill's subsequent advocacy of mental experimentation.

In the second place, Mill is entirely precluded from claiming the mathematical sciences as peculiar in their method, because one of the principal points of his philosophy is to show that they are not peculiar. It is the outcome of his philosophy to show that they are founded on a directly empirical basis, like the rest of the sciences. He speaks[1] of geometry as a 'strictly physical science,' and

---

[1] Book iii, chap. xxiv, sec. 7, about the tenth line.

asserts that every theorem of geometry is a law of external nature, and might have been ascertained by generalising from observation and experiment.[1] What will our physicists say to a *strictly physical science,* which can be experimented on in the private laboratory of the philosopher's mind? What a convenient science! What a saving of expense in regard of apparatus, and materials, and specimens.[2]

Incidentally, it occurs to me to ask whether Mill, in treating geometry, had not forgotten a little sentence which sums up the conclusion of the first section of his chapter on Names.[3] Here he luminously discusses the question whether names are more properly said to be the names of things, or of our ideas of things. After giving some reasons of apparent cogency, he concludes emphatically in these words: 'Names, therefore, shall always be spoken of in this work as the names of things themselves, and not merely of our ideas of things.' Here is really a difficulty. *Straight line* is certainly a name, and yet it can hardly be the name of a thing which is not a straight line. It must then be the name either of a real straight line, or of our idea of a straight line. But Mill distinctly denied that there were such things as straight lines, 'in our planet at least;' hence the name (unless indeed it be the name of lines in other

---

[1] Same section, beginning of second paragraph.

[2] Since writing the above, I have made the significant discovery that in the first and second editions, a clause follows the passage quoted from Book i, chap. v, sec. 1, paragraph 5 (vol. i, middle of p. 119), in the following words:—'A process by which, I will venture to affirm, not a single truth ever was arrived at, except truths of psychology, a science of which Ideas or Conceptions are avowedly (along with other mental phenomena) the subject matter.' These words do not appear in the fifth and ninth editions. Now, as Mill could not possibly pretend to include geometry, *a strictly physical science,* under psychology, we find him implying or rather asserting, that *not a single truth ever was arrived at* in geometry by the very method of handling our ideas on which he depends for the knowledge of the axioms of geometry. The striking out of these words seems to indicate that he had perceived the absolute conflict of his two doctrines; yet he maintains his opinion about *the cardinal error* of handling ideas, and merely deletes a too glaring inconsistency which results from it.

[3] Book i, chap. ii, sec. 1, near the end.

planets) must be the name of our ideas of straight lines. He promised expressly that names 'in this work,' that is, in the *System of Logic*, should *always* be spoken of as the names of things themselves. It must have been by oversight, then, that he forgot this emphatic promise in a later chapter of the same volume. We may excuse an accidental *lapsus memoriæ*, but a philosopher is unfortunate who makes many such lapses in regard to the fundamental principles of his system.

But let us overlook Mill's breach of promise, and assume that we may properly employ ideal experiments. We are told [1] that, though it is impossible ocularly to follow lines 'in their prolongation to infinity,' yet this is not necessary. 'Without doing so we may know that if they ever do meet, or if, after diverging from one another, they begin again to approach, this must take place not at an infinite, but at a finite distance. Supposing, therefore, such to be the case, we can transport ourselves thither in imagination, and can frame a mental image of the appearance which one or both of the lines must present at that point, which we may rely on as being precisely similar to the reality.' Now, we are also told [2] that 'neither in nature nor in the human mind do there exist any objects exactly corresponding to the definitions of geometry.' Not only are there no perfectly straight lines, but there are not even lines without breadth, Mill says,[3] 'We cannot *conceive* a line without breadth ; we can form no mental picture of such a line: all the lines which we have in our minds are lines possessing breadth.' Now I want to know what Mill means by the prolongation of a line which has thickness and is not straight. Let us examine this question with some degree of care.

In the first place, if the line, instead of being length without breadth, according to Euclid's definition, has thickness, it must be a wire ; if it had had two dimensions without

---

[1] Book ii, chap. v, sec. 5, beginning of fourth paragraph.
[2] Book ii, chap. v, sec. 1, beginning of third paragraph.
[3] Same section, second paragraph, eleven lines from end.

the third, it would surely have been described as a surface, not a line. But then I want to know how we are to understand *the prolongation of a wire*. Is the course of the wire to be defined by its surface or by its central line, or by a line running deviously within it? If we take the last, then, the line being devious and uncertain, its prolongation must be undefined. If we take a certain central line, then either this line has breadth or it has no breadth; if the former, all our difficulties recur; if the latter—— Well, Mill denied that we could form the idea of such a line. The same difficulty applies to any line or lines upon the surface, or to the surface itself regarded as a curved surface without thickness. Unless, then, we can get rid of thickness in some way or other, I feel unable to understand what the prolongation of a line means.

But let us overlook this difficulty, and assume that we have got Euclid's line—length without breadth. In fact, Mill tells us [1] that 'we can reason about a line as if it had no breadth,' because we have 'the power, when a perception is present to our senses, or a conception to our intellects, of *attending* to a part only of that perception or conception, instead of the whole.' I believe that this sentence supplies a good instance of a *non sequitur*, being in conflict with the sentence which immediately follows. Mill holds that we learn the properties of lines by experimentation on ideas in the mind; these ideas must surely be conceived, and they cannot be conceived without thickness. Unless, then, the *reasoning* about a line is quite a different process from *experimenting*, I fail to make the sentences hold together at all. If, on the other hand, we can reason about lines without breadth, but can only experiment on thick lines, would it not be much better to stick to the reasoning process, whatever it may be, and drop the mental experimentation altogether?

But let that pass. Suppose that, in one way or other, we manage to *attend* only to the direction of the line, not its

[1] Same paragraph, seventeen lines from end.

thickness. Now, the line cannot be a straight line, because Mill tells us that neither in nature nor in the human mind is there anything answering to the definitions of geometry, and the second definition of Euclid defines a straight line. If not straight, what is it? Crooked, I presume. What, then, are we to understand by the prolongation of a crooked line? If the crooked line is made up of various portions of line tending in different directions, if, in short, it be a zigzag line, of course we cannot prolong it in all those directions at once, nor even in any two directions, however slightly divergent. Let us adopt, then, the last bit of line as our guide. If this bit be perfectly straight, there is no difficulty in saying what the prolongation will be. But then Mill denied that there could be such a bit of straight line; for the length of the bit could scarcely have any relevance in a question of this sort. If not a straight line, it may yet be a piece of an ellipse, parabola, cycloid, or some other mathematical curve. But if a piece of an ellipse, do we mean a piece of a perfect ellipse? In that case one of the definitions of geometry has something answering to it in the mind at least; and if we conceive the more complicated mathematical curves, surely we can conceive the straight line, the most simple of curves. But if these pieces of line are not perfect curves, that is, do not fulfil definite mathematical laws, what are they? If they also are crooked, and made up of fragments of other lines and curves, all the difficulty comes over again. Apparently, then, we are driven to the conception of a line, no portion of which, however small, follows any definite mathematical law whatever. For if any portion has a definite law, the last portion may as well be supposed to be that portion; then we can prolong it in accordance with that law, and the result is a perfect mathematical line or curve, of which Mill denied the existence *either in nature or in the human mind.* We are driven, then, to the final result that no portion of any line follows any mathematical law whatever. Each line must follow its own sweet will. What then are we to understand by the prolongation of such a line ? Surely the whole thing is reduced to the absurd.

But in this inquiry we must be patient. Let us forget the non-existence of straight lines, the cardinal error of mental experimentation, and whatever little oversights we have yet fallen upon. Let us suppose there really are geometrical figures which we can treat in the manner of 'a strictly physical science,' such as geometry seems to be. What lessons can we draw from Mill's Logic as to the mode of treating the figures ? A plain answer is contained in the following extract from the second volume :—

'Every theorem in geometry,' he says,[1] 'is a law of external nature, and might have been ascertained by generalising from observation and experiment, which in this case resolve themselves into comparison and measurement.'

Here we are plainly told that the solution of *every* theorem in geometry may be accomplished by a process of which measurement is, to say the least, a necessary element. No doubt a good deal turns upon the word 'generalising,' by which I believe Mill to mean that what is true of the figure measured will be true of all like figures in general. Give him, however, the benefit of the doubt, and suppose that, after measuring, we are to apply some process of reasoning before deciding on the properties of our figure. Still it is plain that if our measurements are not accurate we cannot attain to perfect or unlimited accuracy in our results, supposing that they depend upon the data given by measurement. Now, I wish to know how Mill would ascertain by generalising from comparison and measurement that the ratio of the diameter and circumference of a circle is that of 1 to $3\cdot14159265358979323846$. . . .

Some years ago I made an actual trial with a pair of compasses and a sheet of paper to approximate to this ratio, and with the utmost care I could not come nearer than one part in 540. Yet Mr. W. Shanks has given the value of

[1] Book iii, chap. xxiv, sec. 7, beginning of second paragraph.

this ratio to the extent of 707 places of decimals,[1] and it is a question of mere labour of computation to carry it to any greater length. It is obvious that the result does not and cannot depend on measurement at all, or else it would be affected by the inaccuracy of that measurement. It is obviously impossible from inexact physical data to arrive at an exact result, and the computations of Mr. Shanks and other calculators are founded on *a priori* considerations, in fact upon considerations which have no necessary connection with geometry at all. The ratio in question occurs as a natural constant in various branches of mathematics, as for instance in the theory of error, which has no necessary connection with the geometry of the circle.

It is amusing to find, too, that Mill himself happens to speak of this same ratio, in his *Examination of Hamilton*,[2] and he there says, 'This attribute was discovered, and is now known, as a result of reasoning.' He says nothing about measurement and comparison. What has become, in this critical case, of the empirical character of geometry which it was his great object to establish? A few lines further on (p. 372) he says that mathematicians could not have found the ratio in question 'until the long train of difficult reasoning which culminated in the discovery was complete.' Now, we are certainly dealing with a theorem of geometry, and if this could have been solved by comparison and measurement, why did mathematicians resort to this long train of *difficult* reasoning?

I need hardly weary the reader by pointing out that the same is true, not merely of many other geometrical theorems, but of all. That the square on the hypothenuse of a right-angled triangle is exactly equal to the sum of the squares on the other sides; that the area of a cycloid is exactly equal to three times the area of the describing circle; that the surface of a sphere is exactly four times that of any of its great circles; even that the three angles of a plane

---

[1] Proceedings of the Royal Society (1872-73), vol. xxi, p. 319.
[2] Second edition, p. 371.

triangle are exactly equal to two right-angles; these and thousands of other certain mathematical theorems cannot possibly be proved by measurement and comparison. The absolute certainty and accuracy of these truths can only be proved deductively. Reasoning can carry a result to infinity, that is to say, we can see that there is no possible limit theoretically to the endless repetition of a process. Thus it is found in the 117th proposition of Euclid's tenth book, that the side and diagonal of a square are incommensurable. No quantity, however small, can be a sub-multiple of both, or, in other words, their greatest common measure is an infinitely small quantity. It has also been shown that the circumference and diameter of a circle are incommensurable. Such results cannot possibly be due to measurement.

It may be well to remark that the expression 'a false empirical philosophy,' which has been used in this article, is not intended to imply that all empirical philosophy is false. My meaning is that the phase of empirical philosophy upheld by Mill and the well-known members of his school, is false. Experience, no doubt, supplies the materials of our knowledge, but in a far different manner from that expounded by Mill.

Here this inquiry must for the present be interrupted. It has been shown that Mill undertakes to explain the origin of our geometrical knowledge on the ground of his so-called *Empirical Philosophy*, but that at every step he involves himself in inextricable difficulties and self-contradictions. It may be urged, indeed, that the groundwork of geometry is a very slippery subject, and forms a severe test for any kind of philosophy. This may be quite true, but it is no excuse for the way in which Mill has treated the subject; it is one thing to fail in explaining a difficult matter: it is another thing to rush into subjects and offer reckless opinions and arguments, which on minute analysis are found to have no coherence. This is what Mill has done, and he has done it, not in the case of geometry alone, but in almost every other point of logical and metaphysical philosophy treated in his works.

II

ON RESEMBLANCE

IN the previous article on John Stuart Mill's Philosophy, I made the strange assertion that *Mill's mind was essentially illogical.* To those who have long looked upon him as their guide, philosopher, and friend, such a statement must of course have seemed incredible and absurd, and it will require a great body of evidence to convince them that there is any ground for the assertion. My first test of his logicalness was derived from his writings on geometrical science. I showed by carefully authenticated extracts, that Mill had put forth views which necessarily imply the existence of perfectly straight lines; yet he had at the same time distinctly denied the existence of such lines. It was pointed out that he emphatically promised to use names *always* as the names of things, not as the names of our ideas of things; yet, as straight lines in his opinion do not exist, the name straight line is either the name of 'just nothing at all,' as James Mill would have said, or else it is the name of our ideas of what they are. It is by experimenting on these ideal straight lines in the mind that we learn the axioms and theorems of geometry according to Mill; nevertheless Mill had denounced, as *the cardinal error of philosophy*, the handling ideas instead of things, and had, indeed, in the earlier editions of the *System of Logic*, asserted that not a single truth ever had been arrived at by this method, except truths of psychology. Mill asserted that we might experiment on lines in the mind by prolonging them to any required distance; but these lines according to Mill's own statements must have thickness, and on minute inquiry it was found impossible to attach any definite meaning at all to the *prolongation of a thick line*. Finally, it was pointed out that, when Mill incidentally speaks of an important mathematical theorem concerning the ratio of the diameter and circumference of the circle, he abandons his empirical philosophy *pro tempore*, and speaks of the ratio in question as being discovered by a long train of difficult reasoning.

Such is the summary of the first small instalment of my evidence. On some future occasion I shall return to the subject of geometrical reasoning, which is far from being exhausted. It will then be proved that, on the question whether geometry is an inductive or a deductive science, Mill held opinions of every phase; in one part of his writings geometry is strictly inductive; in another part it is improperly called inductive; elsewhere, it is set up as the type of a deductive science, and anon it becomes a matter of direct observation and experiment; presently Mill discovers unexpectedly, that there is no difference at all between an inductive and a deductive science; the true distinction is between a deductive and an experimental science. But Mill characteristically overlooks the fact that if the difference lies between a deductive and an experimental science, and not between a deductive and an inductive science, then a similar line of difference must be drawn between an inductive and an experimental science, although Mill's inductive methods are the Four Experimental Methods.

But the origin of our geometrical knowledge is a very slippery subject, as I before allowed. It would not be fair to condemn Mill for the troubles in which he involved himself in regard to such a subject if there were no other counts proved against him. Certainly, he selected geometry as a critical test of the truth of his empirical philosophy, but he may have erred in judgment in choosing so trying a test. Let us, therefore, leave geometry for the present, and select for treatment in this second article a much broader and simpler question—one which lies at the basis of the philosophy of logic and knowledge. We will endeavour to gain a firm comprehension of Mill's doctrine concerning *the nature and importance of the relation of Resemblance*. This question touches the very nature of knowledge itself. Now, critics who are considered to be quite competent to judge, have declared that Mill's logic is peculiarly distinguished by the thorough analysis which it presents of the cognitive and reasoning processes. Mill has not restricted himself to the empty forms and methods of argument, but has pushed his inquiry, as they think, boldly into the psychology and philosophy of reasoning. In the *System of Logic*, then, we shall find it clearly decided whether resemblance is, or is not, the fundamental relation with which reasoning is concerned. It was the doctrine of Locke, as fully expounded in the fourth book of his great essay, that knowledge is the perception of the agreement or disagreement of our ideas.

'Knowledge then,' says Locke, 'seems to me to be nothing but the perception of the connection and agreement, or disagreement and repugnancy, of any of our ideas. In this alone it consists. Where this perception is, there is knowledge; and where it is not, there, though we may fancy, guess, or believe, yet we always come short of knowledge.'

Many other philosophers have likewise held that a certain agreement between things, variously described as resemblance, similarity, identity, sameness, equality, etc., really constituted the whole of *reasoned knowledge*, as distinguished from the mere knowledge of sense. Condillac adopted this view and stated it with admirable breadth and brevity, saying, ' L'évidence de raison consiste uniquement dans l'identité.'

Mill has not failed to discuss this matter, and his opinion on the subject is most expressly and clearly stated in the chapter upon the Import of Propositions.[1] He analyses the state of mind called Belief, and shows that it involves one

[1] Book i, chap. v.

or more of five matters of fact, namely, Existence, Coexistence, Sequence, Causation, Resemblance. One or other of these is asserted (or denied) in every proposition which is not merely verbal. No doubt relations of the kinds mentioned form a large part of the matter of knowledge, and they must be expressed in propositions in some way or other. I believe that they are expressed in the terms of propositions, while the copula always signifies *agreement*, or, as Condillac would have said, *identity* of the terms. But we need not attempt to settle a question of this difficulty. We are only concerned now with the position in his system which Mill assigns to Resemblance. This comes last in the list, and it is with some expression of doubt that Mill assigns it a place at all. He says [1]—

'Besides propositions which assert a sequence or coexistence between two phenomena, there are therefore also propositions which assert resemblance between them; as, This colour is like that colour;—The heat of to-day is *equal* to the heat of yesterday. It is true that such an assertion might with some plausibility be brought within the description of an affirmation of sequence, by considering it as an assertion that the simultaneous contemplation of the two colours is *followed* by a specific feeling termed the feeling of resemblance. But there would be nothing gained by encumbering ourselves, especially in this place, with a generalisation which may be looked upon as strained. Logic does not undertake to analyse mental facts into their ultimate elements. Resemblance between two phenomena is more intelligible in itself than any explanation could make it, and under any classification must remain specifically distinct from the ordinary cases of sequence and coexistence.'

It would seem, then, that Mill had, to say the least, contemplated the possibility of resolving Resemblance into something simpler, namely, into a special case of sequence and coexistence; but he abstains, not apparently because it would be plainly impossible, but because logic does not undertake ultimate analysis. It would encumber us with a 'strained generalisation,' whatever that may be. He there-

[1] Book i, chap. v, sec. 6.

fore accords it provisionally a place among the matters of fact which logic treats.

Postponing further consideration of this passage, we turn to a later book of the *System of Logic*, in which Mill expresses pretty clearly his opinion, that Resemblance is *a minor kind of relation* to be treated last in the system of Logic, as being of comparatively small importance. In the chapter headed 'Of the remaining Laws of Nature,'[1] we find Mill distinctly stating that[2] 'the propositions which affirm Order in Time, in either of its two modes, Coexistence and Succession, have formed, thus far, the subject of the present Book. And we have now concluded the exposition, so far as it falls within the limits assigned to this work, of the nature of the evidence on which these prepositions rest, and the processes of investigation by which they are ascertained and proved. There remain three classes of facts: Existence, Order in Place, and Resemblance, in regard to which the same questions are now to be resolved.'

From the above passage we should gather that Resemblance has not been the subject treated in the preceding chapters of the third book, or certainly not the chief subject.

Of the remaining three classes of facts, Existence is dismissed very briefly. So far as relates to simple existence, Mill thinks[3] that the inductive logic has no knots to untie, and he proceeds to the remaining two of the great classes into which facts have been divided. His opinion about Resemblance is clearly stated in the second section of the same chapter, as follows—

'Resemblance and its opposite, except in the case in which they assume the names of Equality and Inequality, are seldom regarded as subjects of science; they are supposed to be perceived by simple apprehension; by merely applying our senses or directing our attention to the two objects at once, or in immediate succession.'

---

[1] Book iii, chap. xxiv.   [2] First section, near the beginning.
[3] Same section.

After pointing out that we cannot always bring two things into suitable proximity, he adds—

'The comparison of two things through the intervention of a third thing, when their direct comparison is impossible, is the appropriate scientific process for ascertaining resemblances and dissimilarities, and is the sum total of what Logic has to teach on the subject.

'An undue extension of this remark induced Locke to consider reasoning itself as nothing but the comparison of two ideas through the medium of a third, and knowledge as the perception of the agreement or disagreement of two ideas : doctrines which the Condillac school blindly adopted, without the qualifications and distinctions with which they were studiously guarded by their illustrious author. Where, indeed, the agreement or disagreement (otherwise called resemblance or dissimilarity) of any two things is the very matter to be determined, as is the case particularly in the sciences of quantity and extension ; there the process by which a solution, if not attainable by direct perception, must be indirectly sought, consists in comparing these two things through the medium of a third. But this is far from being true of all inquiries. The knowledge that bodies fall to the ground is not a perception of agreement or disagreement, but of a series of physical occurrences, a succession of sensations. Locke's definitions of knowledge and of reasoning required to be limited to our knowledge of, and reasoning about, Resemblances.'

We learn from these passages, then, that science and knowledge have little to do with resemblances. Except in the case of equality and inequality, *resemblance is seldom regarded as the subject of science,* and Mill apparently accepts what he holds to be the prevailing opinion. The sum total of what logic has to teach on this subject is that two things may be compared through the intervention of a third thing, when their direct comparison is impossible. Locke *unduly* extended this remark when he considered reasoning itself as nothing but the comparison of two ideas through the medium of a third. Locke's definitions of knowledge and of reasoning require to be limited to our knowledge of, and reasoning about, resemblances.

In the preceding part of the third book of the *System*

of *Logic*, then, we have not been concerned with Resemblance. The subjects discussed have been contained in propositions which affirm Order in Time, in either of its modes, Coexistence and Succession. Resemblance is another matter of fact, which has been postponed to the twenty-fourth chapter of the third book, and there dismissed in one short section, as being *seldom regarded as a subject of science*. Under these circumstances we should hardly expect to find that Mill's so-called Experimental Methods are wholly concerned with resemblance. Certainly these celebrated methods are the subject of science; they are, according to Mill, the great methods of scientific discovery and inductive proof; they form the main topic of the third book of the Logic, indeed, they form the central pillars of the whole *System of Logic*. It is a little puzzling, then, to find that the names of these methods seem to refer to Resemblance, or to something which much resembles resemblance. The first is called the Method of Agreement; the second is the Method of Difference; the third is the Joint Method of Agreement and Difference; and the remaining two methods are confessedly developments of these principal methods. Now, does Agreement mean Resemblance or not? If it does, then the whole of the third book may be said to treat of a relation which Mill has professedly postponed to the second section of the twenty-fourth chapter.

Let us see what these methods involve. The canon of the first method is stated in the following words,[1] which many an anxious candidate for academic honours has committed to memory:—

'If two or more instances of the phenomenon under investigation have only one circumstance in common, the circumstance in which alone all the instances agree is the cause (or effect) of the given phenomenon.'

Now, when two or more instances of the phenomenon under investigation agree, do they, or do they not, resemble

[1] Book iii, chap. viii, sec. 1, near the end.

each other ? Is agreement the same relation as resemblance, or is it something different ? If, indeed, it be a separate kind of relation, it must be matter of regret that Mill did not describe this relation of agreement when treating of the 'Import of Propositions.' Surely the propositions in which we record our observations of 'the phenomenon under investigation' must affirm agreement or difference, and as the experimental methods are the all-important instruments of science, these propositions must have corresponding importance. Perhaps, however, we shall derive some light from the context ; reading on a few lines in the description of the Method of Difference,[1] we find Mill saying that

'In the Method of Agreement we endeavoured to obtain instances which agreed in the given circumstance but differed in every other : in the present method (*i.e.* the Method of Difference) we require, on the contrary, two instances resembling one another in every other respect, but differing in the presence or absence of the phenomenon we wish to study.'

It would really seem, then, as if the great Experimental Method depends upon our discovering two instances *resembling* one another. Here resemblance is specified by name. We seem to learn clearly that Agreement must be the same thing as Resemblance ; if so, Difference must be its opposite. Proceeding accordingly to consider the Method of Difference we find its requirements described in these words :[2]—
'The two instances which are to be compared with one another must be exactly similar, in all circumstances except the one which we are attempting to investigate.'

This exact similarity is not actual identity, of course, because the instances are *two*, not *one*. Is it then resemblance ? If so, we again find the principal subject of Mill's logic to be that which he relegated to section 2 of chapter xxiv. If we proceed with our reading of Mill's chapter on the 'Four Experimental Methods,' we still find sentence after sentence dealing with this relation of resemblance,

[1] Same chapter, second section.
[2] Same chapter, third section, third paragraph, fourth line.

sometimes under the very same name, sometimes under the names of similarity, agreement, likeness, etc. As to its apparent opposite, *difference*, it seems to be the theme of the whole chapter. The Method of Difference is that wonderful method which can prove the most general law on the ground of two instances! But of this peculiarity of the Method of Difference I shall treat on another occasion.

Perhaps, however, after all I may be misrepresenting Mill's statements. It crosses my mind that by Resemblance he may mean something different from *exact similarity*. The Methods of Agreement and Difference may require that complete likeness which we should call *identity of quality*. It is only fair to inquire then, whether he uses the word Resemblance in a broad or a narrow sense. On this point Mill leaves us in no doubt; for he says distinctly,[1] 'This resemblance may exist in all conceivable gradations, from perfect undistinguishableness to something extremely slight.'

Again on the next page, while distinguishing carefully between such different things as numerical identity and undistinguishable resemblance, he clearly countenances the wide use of the word resemblance, saying,[2] 'Resemblance, when it exists in the highest degree of all, amounting to undistinguishableness, is often called identity.' It seems then, that all grades of likeness or similarity, from undistinguishable identity down to *something extremely slight*, are all properly comprehended under resemblance; and it is difficult to come to any other conclusion than that the agreement and similarity and difference treated throughout the Experimental Methods are all cases of that minor relation, seldom considered the subject of science, which was postponed by Mill to the second section of the twenty-fourth chapter.

But the fact is that I have only been playing with this matter. I ought to have quoted at once a passage which was in my mind all the time—one from the chapter on the Functions and Value of the Syllogism. Mill sums

[1] Book i, chap. iii, sec. 11, paragraph 4.
[2] Same section, fifth paragraph, third line.

up the conclusion of a long discussion in the following words:[1]—

'We have thus obtained what we were seeking, an universal type of the reasoning process. We find it resolvable in all cases into the following elements: Certain individuals have a given attribute; an individual or individuals resemble the former in certain other attributes; therefore they resemble them also in the given attribute.'

All reasoning, then, is resolvable into a case of resemblance; the word *resemble* is itself used twice over, and, as I shall hereafter show, the word *attribute*, synonymous with property, is but another name, according to Mill, for resemblance. It is true that this quotation is taken from the second book of the *System*, not from the preceding part of the third book to which Mill referred as not having treated of resemblance. But this can hardly matter, as he speaks of the *universal type of the reasoning process*, which must include of course the whole of the inductive methods expounded in the third book.

But in case the reader should not be quite satisfied, I will give yet one more quotation, taken from the twentieth chapter of the third book, a chapter therefore which closely precedes the chapter on 'The Remaining Laws of Nature,' where Mill despatches Resemblance. This chapter treats nominally of analogy, but what must be our surprise to find that in reality it treats from beginning to end of Resemblance! This is the way in which he describes reasoning by analogy [2]—

'It is on the whole more usual, however, to extend the name of analogical evidence to arguments from any sort of resemblance, provided they do not amount to a complete induction: without peculiarly distinguishing resemblance of relations. Analogical reasoning, in this sense, may be reduced to the following formula:—Two things resemble each other in one or more respects; a certain proposition is true of the one; therefore, it is true of the other. But we have nothing here by which to discriminate analogy from induction, since this type will serve for all reasoning from experience. In the strictest induction, equally

---

[1] Book ii, chap. iii, sec. 7, at beginning.
[2] Book iii, chap. xx, beginning of second section.

with the faintest analogy, we conclude because A resembles B in one or more properties, that it does so in a certain other property.'

It seems, then, that the universal type of the reasoning process wholly turns upon the pivot of resemblance. The stone which was despised and slightingly treated in a brief section of the twenty-fourth chapter, has become the corner-stone of Mill's logical edifice. It would almost seem as if Mill were one of those persons who are said to think independently with the two halves of their brain. On the one side of the great longitudinal fissure must be held the doctrine that resemblance is seldom a subject of science; on the other side, Mill must have thought out the important place which resemblance holds as the universal type of the reasoning and inductive processes. Double-mindedness, the Law of Obliviscence, or some *Deus ex machinâ*, must be called in; for it is absurd to contemplate the possibility of reconciling Mill's statement of the *universal type of all reasoning* with his remarks upon Locke's doctrine. Locke, he says in the passage already quoted, *unduly extended the importance of resemblance, when he made all reasoning a case of it,* and Locke's definition of knowledge and of reasoning required *to be limited to our knowledge of and reasoning about resemblances*. Yet, according to Mill himself, the *universal type of* ALL *reasoning* turns wholly on resemblance. Under such circumstances, it is impossible to discuss seriously the value of Mill's analysis of knowledge. Which part of the analysis are we to discuss? That in which resemblance is treated as the basis of all reasoning, or that in which it belongs to the 'remaining' and 'minor matters of fact,' which had not been treated in the books of induction, and which therefore remained to be disposed of?

We have not yet done with this question of resemblance; it is the fundamental question as regards the theory of knowledge and reasoning, and, even at the risk of being very tedious, I must show that in the deep of Mill's inconsistency there is still a lower deep. I have to point out that some of his opinions concerning the import of propositions may be thus formulated:—

1. The names of attributes are names for the resemblances of our sensations.
2. Certain propositions affirm the possession of properties, or attributes, or common peculiarities.
3. Such propositions do not, properly speaking, assert resemblance at all.

Proceeding in the first place to prove that Mill has made statements of the meaning attributed to him, we find the matter of the first in a note [1] written by Mill in answer to Mr. Herbert Spencer, who had charged Mill with confounding exact likeness and literal identity. With the truth of this charge we will not concern ourselves now; we have only to notice the following distinct statement: 'What, then, is the common something which gives a meaning to the general name? Mr. Spencer can only say, it is the similarity of the feelings; and I rejoin, the attribute is precisely that similarity. The names of attributes are in their ultimate analyses names for the resemblances of our sensations (or other feelings). Every general name, whether abstract or concrete, denotes or connotes one or more of those resemblances.' Mill's meaning evidently is that when you apply a general name to a thing, as for instance in calling snow *white*, you mean that there is a resemblance between snow and other things in respect of their whiteness. The general name *white* connotes this resemblance; the abstract name *whiteness* denotes it.

Let us now consider a passage in the chapter on the Import of Propositions, which must be quoted at some length.[2]

'It is sometimes said, that all propositions whatever, of which the predicate is a general name, do, in point of fact, affirm or

---

[1] Book ii, chap. ii, sec. 3, near the beginning of the third paragraph of the footnote. This note does not occur in some of the early editions.

[2] Book i, chap. v, sec. 6, second paragraph.

deny resemblance. All such propositions affirm that a thing belongs to a class; but things being classed together according to their resemblance, everything is of course classed with the things which it is supposed to resemble most; and thence, it may be said, when we affirm that gold is a metal, or that Socrates is a man, the affirmation intended is, that gold resembles other metals, and Socrates other men, more nearly than they resemble the objects contained in any other of the classes co-ordinate with these.'

Of this doctrine Mill goes on to speak in the following curious remarks,[1] to which I particularly invite the reader's attention:—

'There is some slight degree of foundation for this remark, but no more than a slight degree. The arrangement of things into classes, such as the class *metal*, or the class *man*, is grounded indeed on a resemblance among the things which are placed in the same class, but not on a mere general resemblance: the resemblance it is grounded on consists in the possession by all those things, of certain common peculiarities; and those peculiarities it is which the terms connote, and which the propositions consequently assert; not the resemblance. For though when I say, Gold is a metal, I say by implication that if there be any other metals it must resemble them, yet if there were no other metals I might still assert the proposition with the same meaning as at present, namely, that gold has the various properties implied in the word metal; just as it might be said, Christians are men, even if there were no men who were not Christians. Propositions, therefore, in which objects are referred to a class, because they possess the attributes constituting the class, are so far from asserting nothing but resemblance, that they do not, properly speaking, assert resemblance at all.'

I have long wondered at the confusion of ideas which this passage exhibits. We are told that the arrangement of things in a class is founded on a resemblance between the things, but not a 'mere general resemblance,' whatever this may mean. It is grounded on the possession of certain 'common peculiarities.' I pass by the strangeness of this expression; I should have thought that *common peculiarity* is a self-contradictory expression in its own terms; but here

[1] Same section, third paragraph.

it seems to mean merely *attribute* or *quality*. The terms then connote this attribute, not the resemblance. Here we are in direct and absolute conflict with Mill's previous statement that *attribute is precisely that similarity*—that common something—which gives a meaning to the general name, and that the names of attributes are, in their ultimate analysis, *names for the resemblances* of our sensations. Previously he said that 'every general name' connotes one or more of these resemblances; now he says that it is 'these peculiarities' which the terms connote, and which the propositions consequently assert, not the resemblances. But these peculiarities are *common peculiarities*—that is, common qualities or attributes. The self-contradiction is absolute and complete, except, indeed, so far as Mill admits that there is 'some slight degree of foundation' for the remark which he is controverting.

We will afterwards consider what is this *slight degree of foundation*; but proceeding for the present with the interpretation of the remarkable passage quoted, we learn that when I say, 'Gold is a metal,' I may imply that if there are other metals it must resemble them; yet, if there were no other metals, I might still assert that gold has the various properties implied in the word metal. The 'Law of Obliviscence' seems to have been at work here; Mill must have quite forgotten that he was speaking of propositions, 'of which the predicate is a general name,' or the name of a class. Now if, as Mill sometimes holds, a class consists only of the things in it,[1] there must be more metals than gold, else metal would not be a general name. If, as Mill elsewhere says, to the contrary effect, the class may exist whether the things exist or not,[2] we still have him on the other horn of the dilemma; for then the meaning of the general name must consist in its connotation, which consists of attributes, which are but another name for resemblances. Yet, forsooth, the proposition does not properly speaking assert resemblances at all.

[1] *System of Logic*, Book ii, chap. ii, sec. 2, fourth paragraph.
[2] Book i, chap. vii, sec. 1, first paragraph.

The important passage quoted above is, as we might readily expect, inconsistent with various other statements in the *System of Logic*, as for instance most of the seventh section of the chapter on Definition, where we are told [1] that the philosopher, 'only gives the same name to things which resemble one another in the same definite particulars,' and that the inquiry into a definition [2] 'is an inquiry into the resemblances and differences among those things.' Elsewhere we are told [3] that 'the general names given to objects imply attributes, derive their whole meaning from attributes; and are chiefly useful as the language by means of which we predicate the attributes which they connote.' Again, in the chapter on the Requisites of a Philosophical Language, he says [4]—

'Now the meaning (as has so often been explained) of a general connotative name, resides in the connotation; in the attribute on account of which, and to express which, the name is given. Thus, the name animal being given to all things which possess the attributes of sensation and voluntary motion, the word connotes those attributes exclusively, and they constitute the whole of its meaning.'

Now, *the attribute, as we learned at starting, is but another name for a Resemblance, and yet a proposition of which the predicate is a general name, does not properly speaking assert resemblance at all.*

The inconsistency is still more striking when we turn to another work, namely, J. S. Mill's edition of his father's *Analysis of the Human Mind*. Here, in a note [5] on the subject of classification, Mill objects to his father's ultra-nominalist doctrine, that 'men were led to class solely for the purpose of economising in the use of names.' Mill proceeds to remark [6] that 'we could not have dispensed

[1] Book i, chap. viii, sec. 7, paragraph 4, about the seventeenth line. This section is numbered 8 in some of the early editions.
[2] Same section, paragraph 8, line 7.
[3] Book iv, chap. iii, eight lines from end of chapter.
[4] Book iv, chap. iv, sec. 2, second line.
[5] Vol. i, p. 260.
[6] Page 261.

with names to mark the points in which different individuals resemble one another: and these are class-names.' Referring to his father's peculiar expression—'individual qualities,' he remarks very properly—

'It is not *individual* qualities that we ever have occasion to predicate. . . . We never have occasion to predicate of an object the individual and instantaneous impressions which it produces in us. The only meaning of predicating a quality at all, is to affirm a resemblance. When we ascribe a quality to an object, we intend to assert that the object affects us in a manner similar to that in which we are affected by a known class of objects.'

A few lines further down he proceeds—

'Qualities, therefore, cannot be predicated without general names; nor, consequently, without classification. Wherever there is a general name there is a class: classification, and general names, are things exactly coextensive.'

This is, no doubt, quite the true doctrine; but what becomes of the paragraph already quoted, which appeared in eight editions of the *System of Logic,* during Mill's lifetime? In that paragraph he asserted that propositions referring an object to a class because they possess the attributes constituting the class, do not, properly speaking, assert resemblance at all. Now, when commenting on his father's doctrine, Mill says that the *only meaning of predicating a quality at all, is to affirm a resemblance.*

In a later note in the same volume Mill is, if possible, still more explicit in his assertion that the predication of general names is a matter of attributes and resemblances. He begins thus [1]—

'Rejecting the notion that classes and classification would not have existed but for the necessity of economising names, we may say that objects are formed into classes on account of their resemblance.'

On the next page he says, in the most distinct manner—

[1] James Mill's *Analysis of the Human Mind.* New edition, vol. i, p. 288.

'Still, a class-name stands in a very different relation to the definite resemblances which it is intended to mark, from that in which it stands to the various accessory circumstances which may form part of the image it calls up. There are certain attributes common to the entire class, which the class-name was either deliberately selected as a mark of, or, at all events, which guide us in the application of it. These attributes are the real meaning of the class-name—are what we intend to ascribe to an object when we call it by that name.'

There can be no possible mistake about Mill's meaning now. The class-name *is intended to mark definite resemblances*. These resemblances must be the attributes which the class-name was either deliberately selected as a mark of, or which guide us in the application of it. These attributes are the *real meaning* of the class-name—are *what we intend to ascribe to an object*, when we call it by that name. Yet we were told in the passage of the *System of Logic* to which I invited the reader's special attention, that propositions in which objects are referred to a class, because they possess the attributes constituting the class, are so far from asserting nothing but resemblance, that they do not, *properly speaking*, assert resemblance at all. A class-name is now spoken of as *intended to mark definite resemblances*. Previously we were informed that, in saying, 'Gold is a metal,' I do not assert resemblance, forsooth, because there might be no other metal but gold. Yet *metal* is spoken of as a class, so that the word metal is a class-name, and the whole discussion refers to propositions of which the predicates are general names.

The fact is, the passage contains more than one *non-sequitur*; it tacitly assumes that *metal* might continue to be a class-name, while there was only one kind of metal, so that there would be nothing else to resemble. Then there is another *non-sequitur* when Mill proceeds straightway to another example, thus—'just as it might be said, Christians are men, even if there were no men who were not Christians.' The words 'just as' here mean that this example bears out the last; but Christians and men being plural, the predicate *men* is now clearly a class-name, and the meaning is that Christians all resemble each other in the attributes connoted by the class-name *man*. Mill adds, indeed, the words 'even if there were no men who were not Christians.' Here is unquestionable confusion of thought. Man is a class-name and connotes the definite resemblances of the objects in the class, even if the class happens to be coextensive with the class Christians. If I say, 'Men are capable of laughter,' the general predicate 'capable of laughter,' connotes a character in which men resemble each other, even though there be no beings capable of laughter who are not men. Thus, when we closely examine the passage in question, it falls to pieces; it has no logical coherence.[1]

I may remark incidentally that it is strange to meet, in a discussion of the fundamental principles of logic and knowledge, with things which have *a slight degree of foundation*. The elementary principles of a science either are true or are not true. There is no middle term. Degree in such matters is out of place. But in Mill's philosophical works, as I shall have various opportunities to show, there is a tendency to what may be called *philosophical trimming*. Instead of saying outright that a thing is false, he says too frequently that it is 'not strictly true,' as in the case referring to the primary ideas of geometry quoted in my last article. Mill's opinions, in fact, so frequently came into conflict with each other, that he acquired the habit of leaving a little room to spare in each of his principal statements: they required a good deal of fitting together. Now 'the slight degree of foundation' for the remark that propositions, of which the predicate is a general name, do assert resemblance, seems to be explained in the two paragraphs which follow that quoted, and these we will now consider.

Mill proceeds to remark that[2] there is sometimes a con-

[1] In my own opinion, an affirmative proposition asserts resemblance in its highest degree, *i.e.* identity, even when the subject and predicate are singular terms; but to prevent confusion, I argue the question on Mill's assumption that the predicate is a general or class-name.

[2] Book i, chap. v, sec. 6, fourth paragraph.

venience in extending the boundaries of a class so as to include things which possess in a very inferior degree, if in any, some of the characteristic properties of the class, provided that they resemble that class more than any other. He refers to the systems of classification of living things, in which almost every great family of plants or animals has a few anomalous genera or species on its borders, which are admitted by a kind of courtesy. It is evident, however, that a matter of this sort has nothing to do with the fundamental logical question whether propositions assert resemblance or not. This paragraph is due to the ambiguity of the word resemblance, which here seems to mean vague or slight resemblance, as distinguished from that incontestable resemblance which enables us to say that things have the same attribute. In fact, a very careful reader of the sections in which Mill treats of resemblance will find that there is frequent confusion between definite resemblance, and something which Mill variously calls 'mere general resemblance' or 'vague resemblance,' which will usually refer to similarities depending on the degree of qualities, or the forms of objects.

There is, however, a second case bearing out Mill's opinion that there is 'some slight degree of foundation' for the remark that propositions whose predicates are general terms affirm resemblance. This is a matter into which we must inquire with some care, so that I give at full length the paragraph relating to it.[1]

'There is still another exceptional case, in which, though the predicate is the name of a class, yet in predicating it we affirm nothing but resemblance, that class being founded not on resemblance in any given particular, but on general unanalysable resemblance. The classes in question are those into which our simple sensations or rather simple feelings, are divided. Sensations of white, for instance, are classed together, not because we can take them to pieces, and say they are alike in this, and not alike in that, but because we feel them to be alike altogether, though in

---

[1] Book i, chap. v, sec. 6, paragraph 5.

different degrees. When, therefore, I say, The colour I saw yesterday was a white colour, or, The sensation I feel is one of tightness, in both cases the attribute I affirm of the colour or of the sensation is mere resemblance—simple likeness to sensations which I have had before, and which have had those names bestowed upon them. The names of feelings, like other concrete general names, are connotative; but they connote a mere resemblance. When predicated of any individual feeling, the information they convey is that of its likeness to the other feelings which we have been accustomed to call by the same name. Thus much may suffice in illustration of the kind of propositions in which the matter-of-fact asserted (or denied) is simple resemblance.'

Such a paragraph as the above is likely to produce intellectual vertigo in the steadiest thinker. In an offhand manner we are told that *this much may suffice* in illustration of an *exceptional case,* in which resemblance happens to be predicated. This resemblance is mentioned slightingly as *mere* resemblance, or *general unanalysable resemblance.* Yet, when we come to inquire seriously what this resemblance is, we find it to be that primary relation of sensation to sensation, which lies at the basis of all thought and knowledge. Professor Alexander Bain is supposed to be, since Mill's death, a mainstay of the empirical school, and, in his works on Logic, he has unfortunately adopted far too much of Mill's views. But, in Professor Bain's own proper writings, there is a vigour and logical consistency of thought for which it is impossible not to feel the greatest respect.

Now we find Mr. Bain laying down, at the commencement of his writings on the Intellect,[1] that the Primary Attributes of Intellect are (1) Consciousness of Difference, (2) Consciousness of Agreement, and (3) Retentiveness. He goes on to say with admirable clearness that discrimination or feeling of difference is an essential of intelligence.

The beginning of knowledge, or ideas, is the discrimination of one thing from another. As we can neither feel,

[1] *Mental and Moral Science, a Compendium of Psychology and Ethics*, 1868, pp. 82, 83. The same doctrine of the nature of knowledge is stated in the treatise on the Senses and the Intellect, second edition, pp. 325-331; in the *Deductive Logic*, pp. 4, 5, 9, and elsewhere.

nor know, without a transition or change of state,—every feeling, and every cognition, must be viewed as in relation to some other feeling, or cognition. There cannot be a single or absolute cognition. Then, again, Mr. Bain proceeds to say that the conscious state arising from Agreement in the midst of difference is equally marked and equally fundamental—

'Supposing us to experience, for the first time, a certain sensation, as redness; and after being engaged with other sensations, to encounter redness again; we are struck with the feeling of identity or recognition; the old state is recalled at the instance of the new, by the fact of agreement, and we have the sensation of red, together with a new and peculiar consciousness, the consciousness of agreement in diversity. As the diversity is greater, the shock of agreement is more lively.'

Then Professor Bain adds emphatically—

'All knowledge finally resolves itself into differences and agreements. To define anything, as a circle, is to state its agreements with some things (genus) and its difference from other things (differentia).'

Professor Bain then treats as the fundamental act of intellect the recognition of redness as identical with redness previously experienced. This is, changing red for white, exactly the same illustration as Mill used, in the example 'The colour I saw yesterday was a white colour.' Now Mr. Bain says, and says truly, that all knowledge finally resolves itself into differences and agreements. Propositions accordingly, which affirm these elementary relations, must really be the most important of all classes of propositions. They must be the elementary propositions which are presupposed or summed up in more complicated ones. Yet such is the class of propositions which Mill dismisses in an offhand manner in one paragraph, as 'still another exceptional case.'

If we look into the details of Mill's paragraph, perplexity only can be the result. He speaks of 'the class being founded not on resemblance in any given particular, but on *general unanalysable resemblance.*' The classes in question are those into which ' our simple sensations, or rather simple feelings, are divided.' Now, what can he possibly mean by *any given particular?* If the colour I saw yesterday was a white colour, that was the given particular in which resemblance existed. No doubt the resemblance is unanalysable, because analysis has done its best, and the matter refers, Mill states, to a *simple sensation.* When we are dealing with the elements of knowledge, of course analysis is no longer applicable. But I confess myself unable to understand why he calls it *general unanalysable resemblance.* If I understand the matter aright, Mill should have said *specific analysed resemblance.* When one red flower is noticed to resemble another red flower in colour, the general resemblance *has been analysed*, and found to consist in a specific resemblance of colour to colour. If I see an orange, I know it to be an orange, because it resembles similar fruits, which I have often heard so called. In the first instance the resemblance may be to my mind mere general resemblance, that is to say, I may not devote separate attention to the several points of resemblance. But if one asks me why I call it an orange, I must analyse my feeling of resemblance, and I then discover that the colour of the fruit resembles the colour of fruit formerly called oranges, and that in regard to the form, the texture of the surface, the hardness, the smell, and so forth, there are other resemblances. My knowledge, as Professor Bain says, finally resolves itself *into differences and agreements.* But the agreements in question are precisely those resemblances, the base-work of all knowledge, which Mill dismisses as *still another exceptional case.*

There is really no mystery or perplexity in the matter except such as Mill has created by the perversity of his intellect. Mill has made that into a species, which is really the *summum genus* of knowledge. Locke truly pronounced knowledge to consist in the perception of agreement or repugnance of our ideas, and Professor Bain has stated the same view with a force and distinctness which leave nothing to be desired. But Mill, strange to say, has treated this all-fundamental relation among 'the Remaining Laws of Nature,' 'Minor Matters of Fact,' or 'Exceptional Cases.' It is usually impossible to trace the causes which led to Mill's perversities, but, in this important case, it is easy to explain the peculiarity of his views on Resemblance. He was labouring under *hereditary prejudice*. His father, James Mill, in his most acute, but usually wrong-headed book, the *Analysis of the Phenomena of the Human Mind*, had made still more strange mistakes. In several curious passages the son argues that we cannot resolve resemblance into anything simpler. These needless arguments are evidently suggested by parts of the *Analysis* in which the father professed to *resolve resemblances into cases of sequence!*

Thus, when James Mill is discussing[1] the Association of Ideas, he objects to Hume specifying Resemblance as one of the grounds of association. He says—

'Resemblance only remains, as an alleged principle of association, and it is necessary to inquire whether it is included in the laws which have been above expounded. I believe it will be found that we are accustomed to see like things together. When we see a tree, we generally see more trees than one; when we see an ox, we generally see more oxen than one; a sheep, more sheep than one; a man, more men than one. From this observation, I think, we may refer resemblance to the law of frequency, of which it seems to form a particular case.'

I cannot help regarding the misapprehension contained in this passage, as perhaps the most extraordinary one which could be adduced in the whole range of philosophical literature. Resemblance is reduced to a *particular case of the law of frequency*, that is, to the frequent recurrence of the same thing, as when, in place of one man, I see many men. But how do I know that they are men, unless I observe that they resemble each other? It is impossible even to speak of *men* without implying that there are various things called men which resemble each other sufficiently to be classed

[1] *Analysis*, first edition, vol. i, p. 79. Second edition, vol. i, p. 111.

together and called by the same name. Nevertheless, James Mill seems to have been actually under the impression that he had got rid of resemblance!

Later on in the same work,[1] indeed, we have the following statement:—

'It is easy to see, among the principles of association, what particular principle it is, which is mainly concerned in Classification, and by which we are rendered capable of that mighty operation; on which, as its basis, the whole of our intellectual structure is reared. That principle is Resemblance. It seems to be similarity or resemblance which, when we have applied a name to one individual, leads us to apply it to another, and another, till the whole forms an aggregate, connected together by the common relation of every part of the aggregate to one and the same name. Similarity, or Resemblance, we must regard as an Idea familiar and sufficiently understood for the illustration at present required. It will itself be strictly analysed, at a subsequent part of this Inquiry.'

In writing this passage, James Mill seems to have forgotten, quite in the manner of his son, that he had before treated Resemblance as an *alleged* principle of association, and had referred it to a particular case of the law of frequency. Here it reappears as the principle on which the whole of our intellectual structure is reared. It is strange that so important a principle should elsewhere be called an 'alleged principle,' and equally strange that it should afterwards be 'strictly analysed.' Before we get down to the basis of our intellectual structure it might be supposed that analysis had exhausted itself.

James Mill gives no reference to the subsequent part of the inquiry where this analysis is carried out, nor do I find that J. S. Mill, or the other editors of the second edition, have supplied the reference. Doubtless, however, the analysis is given in the second section of Chapter XIV, where, in treating of Relative Terms,[2] he inquires into the

[1] *Analysis,* first edition, vol. i, pp. 212, 213. Second edition, vol. i, pp. 270, 271.

[2] First edition, vol. ii, p. 10. Second edition, vol. ii, pp. 11, 12.

meaning of Same, Different, Like, or Unlike, and comes to the conclusion that the resemblance between sensation and sensation is, after all, only sensation. He says—

'Having *two* sensations, therefore, is not only having sensation, but the only thing which can, in strictness, be called having sensation; and the having two and knowing they are two, which are not two things, but one and the same thing, is not only sensation, and nothing else than sensation, but the only thing which can, in strictness, be called sensation. The having a new sensation, and knowing that it is new, are not two things, but one and the same thing.'

This is, no doubt, a wonderfully acute piece of sophistical reasoning; but I have no need to occupy space in refuting it, because J. S. Mill has already refuted it in several passages which evidently refer to his father's fallacy. Thus, I have already quoted, at the commencement of this article, a statement in which J. S. Mill argues that resemblance between two phenomena is more intelligible than any explanation could make it. Again, in editing his father's *Analysis*, Mill comments at some length upon this section,[1] showing that it does not explain anything, nor leave the likenesses and unlikenesses of our simple feelings less ultimate facts than they were before.

But though Mill thus refuses to dissolve resemblance away altogether, his thoughts were probably warped in youth by the perverse doctrines which his father so unsparingly forced upon his intellect. Too early the brain-fibres received a decided *set*, from which they could not recover, and all the power and acuteness of Mill's intellect were wasted in trying to make things fit, which could not fit, because mistakes had been made in the very commencement of the structure.

This misapprehension of the Mills, *père et fils*, concerning resemblance, is certainly one of the most extraordinary instances of perversity of thought in the history of philosophy. That which is the *summum genus* of reasoned

[1] Vol. ii, pp. 17-20.

knowledge, they have either attempted to dissolve away altogether, or, after grudgingly allowing its existence, have placed in the position of a minor species and exceptional case. Yet it is impossible to use any language at all without implying the relation of resemblance and difference in every term. There is not a sentence in Mill's own works in which this fact might not be made manifest after a little discussion. We cannot employ a general name without implying the resemblance between the significates of that name, and we cannot select any class of objects for attention without discriminating them from other objects in general. To propose *resemblance* itself as the subject of inquiry presupposes that we distinguish it from other possible subjects of inquiry. Thus, when James Mill is engaged (in a passage already quoted) in dissipating the relation of resemblance, he presupposes resemblance in every name. What is a *new* sensation, unless it resembles other *new* sensations in being discriminated from *old* sensations? What is a *sensation* unless it resembles other sensations in being separated in thought from things which are *not-sensations*? But it is truly amusing to find that, in the very first sentence of the paragraph immediately following that quoted, James Mill uses the word resemblance. He says:[1] 'The case between sensation and sensation resembles that between sensation and idea.' Nevertheless, James Mill sums up the result of the section of his work in question by the following:[2]—

'It seems, therefore, to be made clear, that, in applying to the simple sensations and ideas their absolute names, which are names of classes, as red, green, sweet, bitter; and also applying to them names which denote them in pairs, as such and such; there is nothing whatsoever but having the sensations, having the ideas, and making marks for them.'

This sentence, if it means anything, means that our sensations and our ideas have no ties between them except in

[1] *Analysis*, first edition, vol. ii, p. 10. Second edition, vol. ii, p. 12.
[2] *Ibid.* first edition, p. 15. Second edition, p. 17.

the common marks or names applied to them. The connection of resemblance is denied existence. This ultra-nominalism of the father is one of the strangest perversities of thought which could be adduced; and though John Stuart Mill disclaims such an absurd doctrine in an apologetic sort of way, yet he never, as I shall now and again have to show, really shook himself free from the perplexities of thought due to his father's errors.

It may seem to many readers that these are tedious matters to discuss at such length. After all, the Import of Propositions and the Relation of Resemblance are matters which concern metaphysicians only, or those who chop logic. But this is a mistake. A system of philosophy—a school of metaphysical doctrines—is the foundation on which is erected a structure of rules and inferences, touching our interests in the most vital points. John Stuart Mill, in his remarkable *Autobiography*, has expressly stated that a principal object of his *System of Logic* was to overthrow deep-seated prejudices, and to storm the stronghold in which they sheltered themselves. These are his words [1]—

'Whatever may be the practical value of a true philosophy of these matters, it is hardly possible to exaggerate the mischiefs of a false one. The notion that truths external to the mind may be known by intuition or consciousness, independently of observation and experience, is, I am persuaded, in these times, the great intellectual support of false doctrines and bad institutions. By the aid of this theory, every inveterate belief and every intense feeling, of which the origin is not remembered, is enabled to dispense with the obligation of justifying itself by reason, and is erected into its own all-sufficient voucher and justification. There never was such an instrument devised for consecrating all deep-seated prejudices. And the chief strength of this false philosophy in morals, politics, and religion, lies in the appeal which it is accustomed to make to the evidence of mathematics and of the cognate branches of physical science. To expel it from these, is to drive it from its stronghold: and because this had never been effectually done, the intuitive school,

---

[1] *Autobiography*, pp. 225-227.

even after what my father had written in his *Analysis of the Mind*, had in appearance, and as far as published writings were concerned, on the whole the best of the argument. In attempting to clear up the real nature of the evidence of mathematical and physical truths, the *System of Logic* met the intuitive philosophers on ground on which they had previously been deemed unassailable; and gave its own explanation, from experience and association, of that peculiar character of what are called necessary truths, which is adduced as proof that their evidence must come from a deeper source than experience. Whether this has been done effectually, is still *sub judice ;* and even then, to deprive a mode of thought so strongly rooted in human prejudices and partialities, of its mere speculative support, goes but a very little way towards overcoming it; but though only a step, it is a quite indispensable one; for since, after all, prejudice can only be successfully combated by philosophy, no way can really be made against it permanently until it has been shown not to have philosophy on its side.'

This is at least a candid statement of motives, means, and expected results. Whether Mill's exposition of the philosophy of the mathematical sciences is satisfactory or not, we partially inquired in the previous article; and in one place or another the inquiry will be further prosecuted in a pretty exhaustive manner. Mill allowed that the character of his solution was still *sub judice*, and it must remain in that position for some time longer. But of the importance of the matter it is impossible to entertain a doubt. If Mill's own philosophy be yet more false than was, in his opinion, the philosophy which he undertook to destroy, we may well adopt his own estimate of the results. '*Whatever*,' he says, '*may be the practical value of a true philosophy of these matters, it is hardly possible to exaggerate the mischiefs of a false one.*' Intensely believing, as I do, that the philosophy of the Mills, both father and son, is a false one, I claim, almost as a right, the attention of those who have sufficiently studied the matters in dispute to judge the arduous work of criticism which I have felt it my duty to undertake.

III

THE EXPERIMENTAL METHODS

My last article on Mill's Philosophy treated of what ought to be, or rather necessarily is, the basis of all reasoning processes—the Relation of Resemblance. It was shown that Mill first of all dismisses this relation as a minor, or even a doubtful matter of fact, or as 'still another exceptional case'; that he then unintentionally makes it the pivot, nay, almost the substance of the reasoning processes, as treated in the book on Induction; yet that, in a later chapter of that book, he returns to the subject of Resemblance as if it had so far been passed over, and finally comes to the conclusion that Resemblance is seldom regarded as the subject of science.

From the base let us proceed to the pillars of Mill's logical edifice. These are the celebrated Methods of Experimental Inquiry — the Method of Agreement, the Method of Difference, the Method of Residues, and the Method of Concomitant Variations; to which may be added, as a kind of corollary, the Joint Method of Agreement and Difference. Mill's exposition of these methods is considered perhaps the most valuable part of his treatise, and much of the celebrity of the book is due to this part. Many people, indeed, whose reading in logic has not been extensive, think that these are Mill's own methods, that he invented them. Any one at all acquainted with the history of logical science knows, of course, that this is not the case, nor did Mill ever claim that it was. Francis Bacon set forth the methods, excepting perhaps that of Residues, in the second book of the *Novum Organum*, vaguely no doubt but with substantial correctness. Taking the nature of heat to exemplify the mode of investigation, he firstly enumerated 'Instances agreeing in the Nature of Heat,' nearly if not exactly corresponding to the Method of Agreement. Next came 'Instances in proximity wanting the Nature of Heat,' by means of which the Method of Difference, or the Joint Method of Agreement and Difference, were brought into play. The Table of Degree or Comparison in Heat forms a rude application of the Method of Concomitant Variations.

Sir John Herschel, again, described these methods with great clearness, and in a manner which I have always preferred to that of Mill. Three of the methods are stated on pp. 151, 152 of his admirable *Discourse on the Study of Natural Philosophy*, and the Method of Residues is given on p. 156. Mill has amply acknowledged his indebtedness to Herschel in several places and ways, and there is not the slightest fault to find with him in that respect. The question is whether Mill, in adopting and formulating the methods anew, and incorporating them into his supposed system of logic, has done better than his predecessors. I shall proceed to show that this is not the case; on the contrary, he has misinterpreted both the foundation and the results of these methods. On some other occasion I shall have to point out that in treating them he has positively confused together an experiment, which is a material operation, with the generalisation by which we pass from the results of the experiment to a general law founded upon it. This confusion of ideas has led him [1] to the astounding and absurd statement that two instances of any phenomenon, treated in strict accordance with the Method of Difference, are sufficient to give with certainty a general law. But, on the present occasion, I treat of the manner

[1] Book iii, chap. x, sec. 2, third and fifth paragraphs. Also chap. xxi, first paragraph.

in which these methods are set up. We must inquire what is the warrant for their validity, and it will be my duty to prove that in this point Mill has fallen into a complete *circulus in probando*. These methods are the only means of proving the connection of cause and effect; yet the methods depend for their validity upon our assurance of the certainty and universality of that connection, that is, upon the universal law of causation.

To students of Mill's logic it is so familiarly known that he bases induction upon the notion of causation, that it may seem superfluous to prove the position. I must nevertheless refer to the chapter treating 'Of the Law of Universal Causation,'[1] where he speaks of 'the notion of Cause being the root of the whole theory of Induction.' Observe the comprehensive force of the expression, 'the whole theory.' Elsewhere,[2] the 'universal fact' of the uniform course of nature is parenthetically described as 'our warrant for all inference from experience,' again an unlimited and most comprehensive remark. The fourth paragraph of the same chapter commences thus: 'Whatever be the most proper mode of expressing it, the proposition that the course of nature is uniform, is the fundamental principle, or general axiom, of Induction.' It is true that Mill sometimes distinguishes between the Uniformity of Nature and the Law of Causation, and gets into perplexities which I have not space to unravel here. It will therefore be better to refer to a later chapter,[3] where Mill places the matter beyond doubt, saying—

'As we recognised in the commencement, and have been enabled to see more clearly in the progress of the investigation, the basis of all these logical operations is the law of causation. The validity of all the Inductive Methods depends on the

---

[1] Chap. v, beginning of second section. As almost all the quotations in this article are taken from the third book of the *System of Logic*, it will be unnecessary again to cite the number of the book, which, unless otherwise specified, will always be the *Third Book*, treating 'Of Induction.'

[2] Chap. iii, beginning of third paragraph.

[3] Chap. xxi, first paragraph.

assumption that every event, or the beginning of every phenomenon, must have some cause ; some antecedent, on the existence of which it is invariably and unconditionally consequent. In the Method of Agreement this is obvious ; that Method avowedly proceeding on the supposition, that we have found the true cause as soon as we have negatived every other. The assertion is equally true of the Method of Difference. That method authorises us to infer a general law from two instances ;[1] one, in which A exists together with a multitude of other circumstances, and B follows ; another, in which, A being removed, and all other circumstances remaining the same, B is prevented. What, however, does this prove ? It proves that B, in the particular instance, cannot have had any other cause than A ; but to conclude from this that A was the cause, or that A will on other occasions be followed by B, is only allowable on the assumption that B must have some cause ; that among its antecedents in any single instance in which it occurs, there must be one which has the capacity of producing it at other times. This being admitted, it is seen that in the case in question that antecedent can be no other than A ; but, that if it be no other than A, it must be A, is not proved, by these instances at least, but taken for granted. There is no need to spend time in proving that the same thing is true of the other Inductive Methods. The universality of the law of Causation is assumed in them all.'

It would be easy to show that this passage is in substance all wrong and unscientific. The idea that we must assume each phenomenon to have one antecedent, and only one, which has the capacity of producing it at (all ?) other times, is quite inconsistent with the scientific idea of causation, as well as with Mill's own statements in other places. It is to a conjunction of causes, joined to all kinds of negative conditions—that is to say, the absence of counteracting causes—that the production of an effect is due ; and this fact alone is enough to disperse Mill's extraordinary assertion that two instances can prove a general law. But the point with which we are concerned now is the complete dependence of the Inductive Methods on the Law of Causation ; not merely the occasional truth of that law, but its *Universality* is assumed in all the methods.

[1] This is the absurd statement alluded to on the preceding page.

The four great pillars of Mill's logical edifice rest, then, upon the universal law of causation. Upon what does this law rest? An ancient system of cosmogony represented the world as resting on an elephant, and the elephant on a tortoise; we want something to correspond to the tortoise. Now it is quite certain that Mill would not derive the law of causation from intuition, consciousness, or any manner of innate source. It was the avowed purpose of his *System of Logic* to show that an appeal to intuition, independently of observation and experience, was the great intellectual support of false doctrines and bad institutions. It is from experience, then, that we must learn the universality of the law of causation. But here the great difficulty of Mill's position begins to be felt. He allows that we do not see this law of nature writ up in plain figures, neither in material nature nor in the mind. The law was quite unknown, he admits, in the earliest ages. It is an induction by no means of the most obvious kind. But Mill's own words must be carefully quoted. Speaking of the fundamental principle, or general axiom of induction, he says [1]—

'I hold it to be itself an instance of induction, and induction by no means of the most obvious kind. Far from being the first induction we make, it is one of the last, or at all events one of those which are latest in attaining strict philosophical accuracy.'

But here comes the rub. If the inductive method, by which we ascertain the connection of causes and effects, presuppose the general law of causation, and this law of causation is one of the latest results of inductive inquiry, how could we ever begin? The experimental methods are of no validity, until we have proved a most general, in fact an *universal* law, which can only be proved by those methods. Logic, let it be always remembered, is, according to Mill, the Science of Proof, and, in such a matter, as the methods of inductive proof, we cannot be supposed to deal with mere surmise. We have now got into this position.

[1] Chap. iii, fourth paragraph.

THE EXPERIMENTAL METHODS

The universal law of causation is represented by the world resting upon the elephant, that is, upon inductive inquiry, and the four legs of that quadruped may correspond to the four pillars of Mill's edifice, the four celebrated Experimental Methods. But upon what do the elephant's legs rest? Upon the world—the world which is already resting on the elephant's back.

To leave the difficulty at this point, and to imply that Mill was totally unconscious of the apparent *circulus in probando*, would be to do him injustice. This case is one of peculiar interest, because it seems to be almost the only case in which Mill was aware of the difficulty from the first, and strove to explain it away. The explanation occurs in the twenty-first chapter of the third book, treating 'Of the Evidence of the Law of Universal Causation.' The substance of the explanation is found even in the first edition; but Mill appeared to feel its inadequacy, and developed his argument in the third, and in some subsequent editions. The result is a notable piece of sophistical reasoning,[1] as follows—

'As was observed in a former place (*supra*, Book iii, chap. iii, sec. 1), the belief we entertain in the universality, throughout nature, of the law of cause and effect, is itself an instance of induction; and by no means one of the earliest which any of us, or which mankind in general, can have made. We arrive at this universal law, by generalisation from many laws of inferior generality. We should never have had the notion of causation (in the philosophical meaning of the term) as a condition of all phenomena, unless many cases of causation, or in other words, many partial uniformities of sequence, had previously become familiar. The more obvious of the particular uniformities suggest, and give evidence of, the general uniformity, and the general uniformity once established enables us to prove the remainder of the particular uniformities of which it is made up. As, however, all rigorous processes of induction presuppose the general uniformity, our knowledge of the particular uniformities from which it was first inferred was not, of course, derived from rigorous induction, but from the loose and uncertain mode of induction

---

[1] Chap. xxi, sec. 2.

per enumerationem simplicem; and the law of universal causation, being collected from results so obtained, cannot itself rest on any better foundation.

'It would seem, therefore, that induction *per enumerationem simplicem* not only is not necessarily an illicit logical process, but is in reality the only kind of induction possible; since the more elaborate process depends for its validity on a law, itself obtained in that inartificial mode. Is there not then an inconsistency in contrasting the looseness of one method with the rigidity of another, when that other is indebted to the looser method for its own foundation?

'The inconsistency, however, is only apparent. Assuredly, if induction by simple enumeration were an invalid process, no process grounded on it could be valid; just as no reliance could be placed on telescopes, if we could not trust our eyes. But though a valid process, it is a fallible one, and fallible in very different degrees: if therefore we can substitute for the more fallible forms of the process, an operation grounded on the same process in a less fallible form, we shall have effected a very material improvement. And this is what scientific induction does.'

Various reflections are suggested by this unfortunate passage. Mill here discovers that the law of causation could not have been derived from rigid induction; he even inserts the words 'of course,' as if no one could have failed to see this. It must therefore be derived from 'the loose and uncertain mode of induction,' with which we shall have more to do. But, in the first place, this treatment of the matter does not square with that in Chapter III, where he treats of the same subject—'The Ground of Induction.' Here he told us, as already quoted, that the uniformity of the course of nature is 'our warrant for all inferences from experience.' Now even 'a loose and uncertain mode of induction' must be a case of inference from experience. Again, Mill distinctly says:[1] 'The statement that the uniformity of the course of nature is the ultimate major premise in all cases of induction, may be thought to require some explanation.' Here he speaks without qualification

[1] Chap. iii, beginning of fifth paragraph.

of 'all cases of induction,' which must include even the loose induction of the ancients. In writing this chapter Mill had not yet discovered that, as induction is based upon causation, causation would have to be based upon something else. Accordingly, though in the third paragraph of the second section of the chapter he mentions the 'loose' induction of the ancients, it is only to depreciate and almost deride it. He thinks it was above all by pointing out the insufficiency of this rude and loose conception of induction, that Bacon merited the title so generally awarded to him of Founder of the Inductive Philosophy.[1] It is curious, then, that Mill in the later chapter finds it necessary to make this loose, uncertain, and insufficient method the basis of his system, inasmuch as it is represented to be our means of learning the universality of the law of causation, on which the validity of the rigid inductive processes depends. Now, in a footnote to Chapter III we are referred to Chapters XXI and XXII; and in Chapter XXI we are similarly referred back to Chapter III. Nevertheless, as I have said, the doctrine of the early chapter fails to square with that of the later one. But there is so much else to come, that I need not dwell upon this discrepancy.

The next reflection that suggests itself is the apparent incongruity of basing the whole of our inductive knowledge of nature upon *a loose and uncertain and insufficient kind of induction*. In several places Mill speaks of this kind of induction with unmitigated scorn. He says [2]—

'The Induction of the ancients has been well described by Bacon, under the name of "Inductio per enumerationem simplicem, ubi non reperitur instantia contradictoria." It consists in ascribing the character of general truths to all propositions which are true in every instance that we happen to know of. This is the kind of induction[3] which is natural to the mind when unaccustomed

---

[1] Same section, fifth paragraph.
[2] Chap. iii, sec. 2, third paragraph.
[3] In the first and second editions we here find the significant words 'if it deserves the name,' that is, of induction; thus we find the great em-

to scientific methods. The tendency, which some call an instinct, and which others account for by association, to infer the future from the past, the known from the unknown, is simply a habit of expecting that what has been found true once or several times, and never yet found false, will be found true again. . . .

'Popular notions are usually founded on induction by simple enumeration; in science it carries us but a little way. We are forced to begin with it; we must often rely on it provisionally, in the absence of means of more searching investigation. But, for the accurate study of nature, we require a surer and a more potent instrument.'

He proceeds, in the next paragraph, still more strongly to denounce this loose method of induction. Speaking of moral and political inquiries, he says—

'The current and approved modes of reasoning on these subjects are still of the same vicious description against which Bacon protested; the method almost exclusively employed by those professing to treat such matters inductively, is the very *inductio per enumerationem simplicem* which he condemns; and the experience which we hear so confidently appealed to by all sects, parties, and interests, is still, in his own emphatic words, *mera palpatio*.'

An obvious difficulty presents itself; if rigid induction depends upon the experimental methods; if these depend upon the law of causation, and this law depends upon *inductio per enumerationem simplicem*; then *the validity of all our inductions depends on a loose and uncertain foundation*. The upper parts of the logical edifice cannot be firmer than its base. Mill, when he comes to the point, shows a creditable consciousness of this difficulty, and accordingly discovers for the occasion that this loose method of induction is not always loose. In the third chapter[1] he remarks—

pirical philosopher, whose work it was to show the inductive basis of all mathematical and other science, accidentally questioning the propriety of allowing the name 'induction' to that process upon which he ultimately bases our knowledge of the universal law of causation, as well as the axioms of geometry. When he inserted these unlucky words he must have forgotten that it was the basis of his system, or else he had not yet discovered the fact.

[1] Sec. 2, fourth paragraph.

'Before we can be at liberty to conclude that something is universally true because we have never known an instance to the contrary, we must have reason to believe that if there were in nature any instances to the contrary, we should have known of them. This assurance, in the great majority of cases, we cannot have, or can have only in a very moderate degree. The possibility of having it, is the foundation on which we shall see hereafter that induction by simple enumeration may in some remarkable cases amount practically to proof.'

Then he refers to the twenty-first chapter, of which the most important passage has already been quoted. Mill allows that there is an apparent inconsistency, but asserts that it is only apparent. The precariousness of the method of simple enumeration is in an inverse ratio to the largeness of the generalisation. As the sphere widens, this unscientific method becomes less and less liable to mislead; and the most universal classes of truths—the law of causation, for instance, and the principles of number and of geometry—are duly and satisfactorily proved by that method alone; nor are they susceptible of any other proof.[1] This is Mill's position, when driven to find a basis for his system.

But then, why does Mill denounce this inductive process as loose, and uncertain, and insufficient, if it is really, as now appears, the basis of all certainty in induction? How can that be *unscientific* upon which all science rests? Why make the whole treatment paradoxical by such a sentence[2] as this? 'For the justification of the scientific method of

[1] Chap. xxi, sec. 3, at beginning, in the third and subsequent editions only.

[2] Same chapter, fourth section. In revising this article I discover that this truly paradoxical statement does not appear in the earlier editions of the *System of Logic*, having been first introduced in the third edition. Later on it disappears again, and in the seventh and subsequent editions, the section commences as follows:—'The assertion, that our inductive processes assume the law of causation, while the law of causation is itself a case of induction, is a paradox only on the old theory of reasoning, which supposes the universal truth, or major premise, in a ratiocination, to be the real proof of the particular truths which are ostensibly inferred from it.' Here Mill slides into a different position; but did space admit, it could be made apparent that his theory of the syllogism quite excludes him from making the universal law of causation the warrant for inductive processes. According to Mill, the evidence

induction as against the unscientific, notwithstanding that the scientific ultimately rests on the unscientific, the preceding considerations may suffice.'

But Mill, though he appears to have explained the inconsistency successfully, has not really cleared himself. He is yet in a coil of difficulties. I now want to know precisely what this loose kind of induction is. Logic, as Mill clearly stated in his Introduction, is the Science of Proof. In so far as belief professes to be founded on proof, the office of logic is to supply a test for ascertaining whether or not the belief is well grounded. The purpose of Mill's treatise is thus concisely set forth [1]—

'Our object, then, will be, to attempt a correct analysis of the intellectual process called Reasoning or Inference, and of such other mental operations as are intended to facilitate this : as well as, on the foundation of this analysis, and *pari passu* with it, to bring together or frame a set of rules or canons for testing the sufficiency of any given evidence to prove any given proposition.'

Now I want to know where, in Mill's treatise, is to be found the analysis of this process of induction *per enumerationem simplicem*? And where is the set of rules and canons for performing it? On this process, as we have found, ultimately rests the proof of all truths, both of mathematical science, and of causation; whatever we prove by the four experimental methods is really proved by the underlying inductive process on which their validity depends. Mill's logic is supposed to present the most thorough analysis of the foundations of our knowledge, and he himself put it forth professedly as intended to clear up the real nature of the evidence of mathematical and physical truths.[2]

for a general truth is resolvable into the particular ones on which it is founded, so that Mill's new position amounts to saying that certain past acts of induction are a warrant for future acts. But where was the warrant for the past acts? It is absolutely impossible to meet all Mill's arguments, because, as each new difficulty presents itself, he invents a new explanation, regardless or rather oblivious, of consistency with his old ones.

[1] Introduction, sec. 7, second paragraph.
[2] Autobiography, p. 226, quoted above, pp. 248, 249.

It was above all things necessary that Mill should have analysed and described this process of 'simple enumeration' with care and completeness, because it is the basis of his whole empirical system. Where is the analysis? Where are the rules of the method? If we search the treatise, we find the process mentioned here and there, but, strange to say, almost always in a depreciatory and scornful manner.

It is a loose, rude, uncertain, insufficient, fallible, unscientific, precarious,—even a *vicious* process. Such are the epithets which Mill applies to the basis of his empirical philosophy, except in the section or two written when he happened to remember that it was the basis. Then, again, where are the rules of this method of induction? If it be usually so insufficient and fallible a support, surely it was all the more requisite that we should have precise rules whereby to judge when it is precarious and when it is not. But the rules and canons given in the treatise are those of the four Experimental Methods, and these rules cannot possibly help us, because the methods themselves derive their validity from the underlying law of causation, which is established by *inductio per enumerationem simplicem*. I say, then, that just where Mill's analysis should have been most careful, and his canons most explicit, there is nothing of the sort, and if we seek for a description of this fundamental kind of inductive reasoning, we find it called by a Latin phrase, and treated with impatience and contempt.

But let us make the best of such descriptions of the fundamental process of his 'system' as Mill favours us with. I have already quoted one passage in which he says that the kind of induction in question 'consists in ascribing the character of general truths to all propositions which are true in every instance that we happen to know of.'

Elsewhere Mill,[1] in reference to coexistences independent of causation, says—

'In the absence, then, of any universal law of coexistence, similar to the universal law of causation which regulates sequence,

---
[1] Chap. xxii, sec. 4, last paragraph.

we are thrown back upon the unscientific induction of the ancients, *per enumerationem simplicem, ubi non reperitur instantia contradictoria*. The reason we have for believing that all crows are black, is simply that we have seen and heard of many black crows, and never one of any other colour. It remains to be considered how far this evidence can reach, and how we are to measure its strength in any given case.'

It is true that in the sections which follow we have some vague discussions as to the circumstances under which we may trust an empirical induction. But in writing these sections Mill seems again to have forgotten that the law of causation is itself founded on the same basis. In the passage quoted above we are told that in the absence of a universal law similar to the universal law of causation, we are thrown back upon the unscientific induction of the ancients. But surely in the case of causation also we are similarly thrown back on this unscientific induction, if we wish to know the ultimate warrant for our inferences. In these sections Mill professes to treat only 'of Coexistences independent of Causation,' such being the title and subject of the whole chapter. He gives no indication how we are to apply the same process to prove the law of causation itself, which is always by him sharply distinguished from the cases treated in the chapter named. In fact, he tells us in the first paragraph of the fourth section, that the application of a system of rigorous scientific induction is precluded in the cases here treated. 'The basis of such a system is wanting: there is no general axiom standing in the same relation to the uniformities of coexistence as the law of causation does to those of succession.' In fact, Mill writes throughout this chapter as if the law of causation had nothing to do with induction by simple enumeration, upon which we are thrown back in other cases.

Turning again then to the most distinct account which we get of this method, we find that induction by simple enumeration consists in ascribing the character of general truths to all propositions which are true in every instance that we happen to know of. Now, the universal law of causation is to the effect that every phenomenon is invariably sequent upon some other phenomenon called the cause. It is the law of *invariable* (and as he sometimes insists) *unconditional sequence*. If we learn the truth of this law by simple enumeration, we must ascribe the character of a general truth to it, because we know it to be true in every instance that we happen to know of. That is to say, in the case of every particular phenomenon which has occurred to us, we must have assured ourselves that there was a cause upon which it was invariably sequent, before we could have the materials for an induction by simple enumeration. The inductive process here, as far as we can gather, consists only in inferring of all cases what we know to be true *without exception* of those which have attracted our attention. But at this point difficulties crowd upon us. Mill can never have formed any clear idea in his mind of the way in which this simple enumeration helps us to the law of causation. The first question to which he supplies no answer is, How in any particular case we know that a phenomenon has a cause, we being supposed ignorant of the universal law of causation? When leading up to his great experimental methods, Mill excites our interest by showing the extreme difficulty of discovering the relation of cause and effect. He says [1]—

'The order of nature, as perceived at a first glance, presents at every instant a chaos followed by another chaos. We must decompose each chaos into single facts. We must learn to see in the chaotic antecedent a multitude of distinct antecedents, in the chaotic consequent a multitude of distinct consequents. This, supposing it done, will not of itself tell us on which of the antecedents each consequent is invariably attendant. To determine that point, we must endeavour to effect a separation of the facts from one another, not in our minds only, but in nature.'

Continuing at the commencement of the next section, we read—

[1] Chap. vii, second paragraph.

'The different antecedents and consequents, being, then, supposed to be, so far as the case requires, ascertained and discriminated from one another; we are to inquire which is connected with which. In every instance which comes under our observation, there are many antecedents and many consequents. If those antecedents could not be severed from one another except in thought, or if those consequents never were found apart, it would be impossible for us to distinguish (*à posteriori* at least) the real laws, or to assign to any cause its effect, or to any effect its cause.'

He goes on to explain that, to effect this analysis, we must be able to meet with some of the antecedents apart from the rest, and observe what follows from them. We must follow the Baconian rule of *varying the circumstances*, and it is this rule which is developed into the four Experimental Methods. But here we are in a most palpable difficulty. We cannot assign any sequent to a particular antecedent without going through the elaborate investigations referred to above, without in fact employing the experimental methods, explicitly or implicitly. Even the first of those methods, that of Agreement, is insufficient for the purpose, because we are told[1] that it has the defect of not proving causation, and can, therefore, only be employed for the ascertainment of empirical laws. *Now we are in a perfect vicious circle.* Causation is proved only by the Method of Difference. That Method derives its validity from the universality of the law of causation. The universality of this law is ascertained by induction by simple enumeration, which requires that we shall have ascertained the truth of the law in every particular case, a thing which clearly could not be done without the Method of Difference. The whole edifice of Mill's inductive logic, elaborately described in the twenty-five long chapters of the third book, collapses. The basis disappears altogether, and the four pillars, the four Experimental Methods, are left supporting themselves in a logical void.

If another stroke were needed for the overthrow of Mill's

[1] Chap. xvii, second paragraph.

vaunted system, it could easily be given in the form of one question. Are there really no apparent exceptions to the universal law of causation? Mill grounds this law upon induction by simple enumeration, *ubi non reperitur instantia contradictoria,* 'when no contradictory instance is encountered.' Applied to the case of causation, this process would require that in all our experience we had never noticed, or at least investigated, a case without ascertaining that the law of causation was verified. What a monstrous assumption is this! Will any one deny that there are whole regions of facts familiarly known to us where we cannot detect the action of causation? What determines the sex of young animals? What produces unexpected forms and diseases, monstrous births, *lusus naturæ*, as they are significantly called? All kinds of tumours, ulcers, and local diseases, spring up in various parts of the human body, and medical science can usually give no explanation of them. It is astonishing how statements made in a work of repute are allowed to pass unquestioned, although directly contrary to the most obvious facts. Of course we may expect or believe that all such phenomena will sooner or later be explained as the effects of undiscovered causes, but such expectation must be *à priori* in its origin, if Mill's own account of the way in which we ascertain the law of causation empirically is true. It is useless to say that we can *prove* the law of causation empirically, when apparent exceptions to its truth are endless in number. A certain probability no doubt may be given to the law empirically, but this does not help Mill, who frequently implies that the law of causation is *certainly and universally true*, and that, as soon as the principle of causation makes its appearance, the precarious inferences derived from simple enumeration are superseded and disappear from the field.[1]

No doubt a skilful controversialist might find in Mill's book many openings for a plausible reply, but none of them would bear cross-examination. It might be pointed out

[1] Chap. xviii, sec. 4, end of third paragraph.

that Mill, in the twenty-first chapter, shows his consciousness of the precarious nature of induction by simple enumeration, but urges that, by widening the sphere of induction, we may indefinitely increase the certainty of the inference. This argument, however, would show complete misapprehension of the theory of probability, and the principles of evidence. Mill, though urging this view of the matter, had never formed any clear ideas on the subject. Observe that the method of simple enumeration consists in ascribing the character of general truths to all propositions which are true in every instance that we happen to know of. Uncertainty enters in a double manner: there is the uncertainty whether what is true of certain particular cases is true of all other cases. This uncertainty would be gradually removed by increasing the number of cases examined. The other uncertainty depends upon the difficulty of showing that any one consequent follows from an invariable antecedent. Now Mill makes it abundantly plain that only the Method of Difference can establish this fact of sequence with certainty. It follows that every other method of ascertaining the connection can give it only with a degree of probability. Then every particular case to which we apply simple enumeration is more or less uncertain; and so far as this uncertainty is common to all the cases, no multiplication of new cases can remove the uncertainty. In fact it comes to this, that the degree of certainty (that is, more properly speaking, probability) which we can give to the universal law of causation cannot exceed, and may fall short of, the certainty of the process by which we discover the connection of causes and effects, prior to the establishment of the great experimental methods. Now as those methods are expounded as the modes of discovering causation, and are often described as the *only modes of proving causation*, we are again left by Mill without any analysis of the real base of his system. There is an evident vicious circle. The Method of Difference proves causation; it reposes on the universal law of causation; which is gathered by simple enumeration from particular cases of causation; which are proved by a process left quite undescribed by Mill, unless it be the Method of Difference. Observe, carefully, that the particular cases to which we apply induction by simple enumeration must be *proved*, or else the uncertainty attaching to all of them will attach also to the universal law, and to the methods of experiment founded upon it.

There is no difficulty in pointing out the mistake which led Mill into so much trouble. That mistake consisted in basing his experimental methods upon the law of causation. His exposition of those methods is faulty and objectionable in several different ways, and, as I have before remarked, the brief and simple rules of Herschel are to be preferred. But the principal fault is, that instead of employing the methods of induction to ascertain the general law of causation, he put the cart before the horse, and used the law of causation to support the methods. He was wrong again in excluding from use the theory of probabilities; he holds that a single perfect experiment *proves* a general law, which must mean, if it means anything, that it proves the law with certainty, a result opposed to all science and to all common sense. It is quite to be expected that a philosopher who seriously proposed to base 'the scientific upon the unscientific,' should meet with paradox and inconsistency in all directions. Such will also be the fate of all who try to uphold Mill's views of the relation between causation and induction.

IV

UTILITARIANISM

IN some respects Mill's Essays, published under the title 'Utilitarianism,' are among his best writings. They have, in the first place, the excellence of brevity. Ninety-six pages, printed in handsome type, make but a light task for the student who wishes to enter into the intricacies of moral doctrine. Moreover, the last Essay consists of a digression concerning the nature and origin of the idea of Justice, and it occupies nearly one-third of the whole book. Thus Mill managed to compress his discussion of so important a subject as the foundations of Moral Right and Wrong into some sixty pleasant pages.

And pleasant pages they certainly are, for they are written in Mill's very best style. Now Mill, even when he is most prolix, when he is pursuing the intricacies of the most involved points of logic and philosophy, can seldom or never be charged with dulness and heaviness. His language is too easy, polished, and apparently lucid. In these Essays on Utilitarianism, he reaches his own highest standard of style. There is hardly any other book in the range of philosophy, so far as my reading has gone, which can be read with less effort. There is something enticing in the easy flow of sentences and ideas, and without apparent difficulty the reader finds himself agreeably borne into the midst of the most profound questions of ethical philosophy, questions which have been the battleground of the human intellect for two thousand five hundred years.

Partly to this excellence of style, partly to Mill's immense reputation, acquired by other works and in other ways, must we attribute the importance which has been generally attached to these ninety-six pages. Probably no other modern work of the same small typographical extent has been equally discussed, criticised, and admired, unless, indeed, it be the Essay on Liberty of the same author. The result is, that Mill has been generally regarded as the latest and best expounder of the great Utilitarian Doctrine—that doctrine which is, by one and no doubt the preponderating school, regarded as the foundation of all moral and legislative progress. Many there are who think that, what Hume and Paley and Jeremy Bentham began, Mill has carried nearly to perfection in these agreeable Essays.

Nothing can be more plain, too, than that Mill himself believed he was dutifully expounding the doctrines of his father, of his father's friend, the great Bentham, and of the other unquestionable Utilitarians among whom he grew up. Mill seems to pride himself upon having been the first, not indeed to invent, but to bring into general acceptance the name of the school to which he supposed himself to belong. He says:[1] 'The author of this essay has reason for believing himself to be the first person who brought the word utilitarian into use. He did not invent it, but adopted it from a passing expression in Mr. Galt's *Annals of the Parish*. After using it as a designation for several years, he and others abandoned it from a growing dislike to anything resembling a badge or watchword of sectarian distinction. But as a name for one single opinion, not a set of opinions—to denote the recognition of utility as a standard, not any particular way of applying it—the term supplies a want in the language, and offers, in

[1] *Utilitarianism*, fifth edition, p. 9, footnote. Except where otherwise specified, the references throughout this article will be to the pages of the fifth edition of *Utilitarianism*.

many cases, a convenient mode of avoiding tiresome circumlocution.'

In the *Autobiography* (p. 79), Mill makes a statement to the same effect, saying—

'I did not invent the word, but found it in one of Galt's novels, the *Annals of the Parish*, in which the Scotch clergyman, of whom the book is a supposed autobiography, is represented as warning his parishioners not to leave the Gospel and become utilitarians. With a boy's fondness for a name and a banner I seized on the word, and for some years called myself and others by it as a sectarian appellation; and it came to be occasionally used by some others holding the opinions it was intended to designate. As those opinions attracted more notice, the term was repeated by strangers and opponents, and got into rather common use just about the time when those who had originally assumed it, laid down that along with other sectarian characteristics.'

It is pointed out, however, by Mr. Sidgwick in his article on Benthamism,[1] that Bentham himself suggested the name 'Utilitarian,' in a letter to Dumont, as far back as June 1802.

Mill explicitly states that it was his purpose in these Essays on Utilitarianism to expound a previously received doctrine of utility. Towards the close of his first chapter, containing General Remarks, he says (p. 6): 'On the present occasion, I shall, without further discussion of the other theories, attempt to contribute something towards the understanding and appreciation of the Utilitarian or Happiness theory, and towards such proof as it is susceptible of.' He proceeds to explain that a preliminary condition of the rational acceptance or rejection of a doctrine is that its formula should be correctly understood. The very imperfect notion ordinarily formed of the Utilitarian formula was the chief obstacle which impeded its reception; the main work to be done, therefore, by a Utilitarian writer was to clear the doctrine from the grosser misconceptions. Thus the question would be greatly simplified, and a large proportion of its difficulties removed. His Essays purport

[1] *Fortnightly Review*, May 1877, vol. xxi, p. 648.

throughout to be a defence and exposition of the Utilitarian doctrine.

But one characteristic of Mill's writings is that there is often a wide gulf between what he intends and what he achieves. There is even a want of security that what he is at any moment urging may not be the logical contrary of what he thinks he is urging. This happens to be palpably the case with the celebrated Essays before us. Mill explains and defends his favourite doctrine with so much affection and so much candour that he finally explains himself into the opposite doctrine. Yet with that simplicity which is a pleasing feature of his personal character, Mill continues to regard himself as a Utilitarian long after he has left the grounds of Paley and Bentham. Lines of logical distinction and questions of logical consistency are of little account to one who cannot distinguish between fact and feeling, between sense and sentiment. It is possible that no small part of the favour with which these Essays have always been received by the general public is due to the happy way in which Mill has combined the bitter and the sweet. The uncompromising rigidity of the Benthamist formulas is softened and toned down. An apparently scientific treatment is combined with so many noble sentiments and high aspirations, that almost any one except a logician may be disarmed.

But nothing can endure if it be not logical. These Essays may be very agreeable reading; they may make readers congratulate themselves on so easily becoming moral philosophers; but they cannot really advance moral science if they represent one thing as being another thing. I make it my business therefore in this article to show that Mill was intellectually unfitted to decide what was utilitarian and what was not. In removing the obstacles to the reception of his favourite doctrine he removed its landmarks too, and confused everything. It is true that I come rather late in the day to show this. Some scores, if not hundreds, of critics have shown the same fact more or less clearly.

Eminent men of the most different schools and tones of thought—such as the Rev. Dr. Martineau, Mr. Sidgwick, Dr. Ward, Professor Birks, the late Professor Grote—have criticised and refuted Mill time after time.

Since commencing my analysis of Mill's Philosophy, I have been surprised to find, too, that some who were supposed to support Mill's school through thick and thin, have long since discovered the inconsistencies which I would now expose at such wearisome length, as if they were new discoveries. Such is the ground which my friend, Professor Croom Robertson, takes in his quarterly review, *Mind*, which must be considered our best authority on philosophical questions. As to this matter of Utilitarianism, a very eminent author, formerly a friend of Mill himself, assures me that the subject is quite threshed out, and implies that there is no need for me to trouble the public any more about it. In fact, it would seem to be allowed within philosophical circles that Mill's works are often wrongheaded and unphilosophical. Yet these works are supposed to have done so much good that obloquy attaches to any one who would seek to diminish the respect paid to them by the public at large. Philosophers, and teachers of the last generation at least, have done their best to give Mill's groundless philosophy a hold upon all the schools and all the press, and yet we of this generation are to wait calmly until this influence dissolves of its own accord. We are to do nothing to lessen the natural respect paid to the memory of the dead, especially of the dead who have unquestionably laboured with single-minded purpose for what they considered the good of their fellow-creatures. But in nothing is it more true than in philosophy, that 'the evil that men do lives after them; the good is oft interred with their bones.' Words and false arguments cannot be recalled. Throw a stone into the surface of the still sea, and you are powerless to prevent the circle of disturbance from spreading more and more widely. True it is, that one disturbance may be overcome and apparently obliterated by other deeper disturbances; but

Mill's works and opinions were disseminated by the immense former influence of the united band of Benthamist philosophers. He is criticised and discussed and repeated, in almost every philosophical work of the last thirty or forty years. He is taken throughout the world as the representative of British philosophy, and it is not sufficient for a few eminent thinkers in Oxford, or Cambridge, or London, or Edinburgh, or Aberdeen, to acknowledge in a tacit sort of way that this doctrine and that doctrine is wrong. Eventually, no doubt, the opinion of the Lecture Halls and Combination Rooms will guide the public opinion; but it may take a generation for tacit opinions to permeate society. We must have them distinctly and boldly expressed. It is especially to be remembered that the public press throughout the English-speaking countries is mostly conducted by men educated in the time when Mill's works were entirely predominant. These men are now for the most part cut off, by geographical or professional obstacles, from the direct influence of Oxford or Cambridge. The circle of disturbance has spread beyond the immediate reach of those centres of thought. To be brief, I do not believe that Mill's immense philosophical influence, founded as it is on confusion of thought, will readily collapse. I fear that it may remain as a permanent obstacle in the way of sound thinking. *Citius emergit veritas ex errore, quam ex confusione.* Had Mill simply erred as did Hobbes about elementary geometry, and Berkeley about infinitesimals, it would be necessary merely to point out the errors and consign them to merciful oblivion. But it is not so easy to consign to oblivion ponderous works so full of confusion of thought that every inexperienced and unwarned reader is sure to lose his way in them, and to take for profound philosophy that which is really a kind of kaleidoscopic presentation of philosophic ideas and phrases, in a succession of various but usually inconsistent combinations. To the public at large, Mill's works still undoubtedly remain as the standard of accurate thinking, and the most esteemed repertory of philosophy. I cannot therefore consider my criticism superfluous, and at the risk of repeating much that has been said by the eminent critics already mentioned, or by others, I must show that Mill has thrown ethical philosophy into confusion as far as could well be done in ninety-six pages.

The nature of the Utilitarian doctrine is explained by Mill with sufficient accuracy in pp. 9 and 10, where he says—

'The creed which accepts as the foundation of morals, Utility, or the Greatest Happiness Principle, holds that actions are right in proportion as they tend to promote happiness, wrong as they tend to produce the reverse of happiness. By happiness is intended pleasure, and the absence of pain; by unhappiness, pain, and the privation of pleasure. To give a clear view of the moral standard set up by the theory, much more requires to be said; in particular, what things it includes in the ideas of pain and pleasure; and to what extent this is left an open question. But these supplementary explanations do not affect the theory of life on which this theory of morality is grounded—namely, that pleasure, and freedom from pain, are the only things desirable as ends; and that all desirable things (which are as numerous in the utilitarian as any other scheme) are desirable either for the pleasure inherent in themselves, or as means to the promotion of pleasure and the prevention of pain.'

Mill proceeds to say that such a theory of life excites inveterate dislike in many minds, and among them some of the most estimable in feeling and purpose. To hold forth no better end than pleasure is felt to be utterly mean and grovelling—a doctrine worthy only of swine. Mill accordingly proceeds to inquire whether there is anything really grovelling in the doctrine—whether, on the contrary, we may not include under pleasure, feelings and motives which are in the highest degree noble and elevating. The whole inquiry turns upon this question—Do pleasures differ in quality as well as in quantity? Can a small amount of pleasure of very elevated character outweigh a large amount of pleasure of low quality? We should never think of estimating pictures by their size. The productions of West and Fuseli, which were the wonder and admiration of our grandparents, can now be bought by the square yard, to cover the bare walls of eating-houses and music-halls. *Sic transit gloria mundi.* But a choice sketch by Turner sometimes sells for many pounds per square inch. It is clear, then, that in the opinion of connoisseurs, which must, for our present purpose, be considered final, high art is almost wholly a matter of quality. Two great pictures by West may be nearly twice as valuable as one; and two equally choice sketches by Turner are twice as good as one; but it would seem hardly possible in the present day for the disciple of 'high art' to bring West and Turner into the same category of thought. I suppose that even Turner will presently begin to wane before 'the higher criticism.'

A corresponding difficulty lies at the very basis of the Utilitarian theory of ethics. The tippler may esteem two pints of beer doubly as much as one; the hero may feel double satisfaction in saving two lives instead of one; but who shall weigh the pleasure of a pint of beer against the pleasure of saving a fellow-creature's life.

Paley, indeed, cut the Gordian knot of this difficulty in a summary manner; he denied altogether that there is any difference between pleasures, except in continuance and intensity. It must have required some moral courage to write the paragraph to be next quoted; yet Paley, however much he may be said to have temporised and equivocated about oaths and subscription to Articles, cannot be accused of want of explicitness in this passage. There is a directness and clear-hitting of the point in Paley's writings which always charms me.

'In strictness, any condition may be denominated happy, in which the amount or aggregate of pleasure exceeds that of pain; and the degree of happiness depends upon the quantity of this excess. And the greatest quantity of it ordinarily attainable in human life, is what we mean by happiness, when we inquire or pronounce what human happiness consists in. In which inquiry I will omit much usual declamation on the dignity and capacity of our nature; the superiority of the soul to the body, of the rational to the animal part of our constitution; upon the worthiness, refinement, and delicacy of some satisfactions, or the meanness, grossness, and sensuality of others; because I hold that pleasures differ in nothing, but in continuance and intensity: from a just computation of which, confirmed by what we observe of the apparent cheerfulness, tranquillity, and contentment, of men of different tastes, tempers, stations, and pursuits, every question concerning human happiness must receive its decision.'[1]

Bentham, it need hardly be said, adopted the same idea as the basis of his ethical and legislative theories. In his uncompromising style he tells us [2] that

'Nature has placed mankind under the governance of two sovereign masters, *pain* and *pleasure*. It is for them alone to point out what we ought to do, as well as to determine what we shall do. On the one hand the standard of right and wrong, on the other the chain of causes and effects, are fastened to their throne. They govern us in all we do, in all we say, in all we think: every effort we can make to throw off our subjection will serve but to demonstrate and confirm it. In words a man may pretend to abjure their empire: but in reality he will remain subject to it all the while. The *principle of utility* recognises this subjection, and assumes it for the foundation of that system, the object of which is to rear the fabric of felicity by the hands of reason and of law. Systems which attempt to question it, deal in sounds instead of sense, in caprice instead of reason, in darkness instead of light.'

Elsewhere Bentham proceeds to show how we may estimate the *values* of pleasures and pains, meaning obviously by *values* the quantities or forces. As these feelings are both the ends and the instruments of the moralist and legislator, it especially behoves us to learn how to estimate these values aright, and Bentham tells us most distinctly.[3]

'To a person, he says, considered *by himself*, the value of a pleasure or pain considered *by itself*, will be greater or less, according to the four following circumstances: (1) Its *intensity*. (2) Its *duration*. (3) Its *certainty* or *uncertainty*. (4) Its *propinquity*

---

[1] *The Principles of Moral and Political Philosophy*, Book i, chap. vi, second paragraph.

[2] *An Introduction to the Principles of Morals and Legislation*, p. 1.

[3] *Principles*, etc. chap. iv, secs. 2-5. The statement is not a verbatim extract but an abridgement of the sections named.

or *remoteness*. But when the value of any pleasure or pain is to be considered for the purpose of estimating the general tendency of the act, we have to take into account also: (5) The *fecundity*, or the chance it has of being followed by sensations of the same kind, that is, pleasures, if it be a pleasure; pains, if it be a pain. (6) Its *purity*, or the chance it has of *not* being followed by sensations of the *opposite* kind: that is, pains, if it be a pleasure; pleasures, if it be a pain. Finally, when we consider the interests of a number of persons, we must also estimate a pleasure or pain with reference to—(7) Its *extent ;* that is, the number of persons to whom it extends, or who are affected by it.'

Thus did Bentham clearly and explicitly lay the foundations of the moral and political sciences, and to impress these fundamental propositions on the memory he framed the following curious mnemonic lines, which may be quoted for the sake of their quaintness :—

> '*Intense, long, certain, speedy, fruitful, pure*—
> Such marks in pleasures and in pains endure.
> Such pleasures seek, if private be thy end :
> If it be public, wide let them *extend*.
> Such *pains* avoid, whichever be thy view :
> If pains *must* come, let them *extend* to few.'

In all that Bentham says about pleasure and pain, there is not a word about the intrinsic superiority of one pleasure to another. He advocates our seeking *pure* pleasures; but with him a pure pleasure was clearly defined as one not likely to be followed by feelings of the opposite kind; the pleasure of opium-eating, for instance, would be called impure, simply because it is likely to lead to bad health and consequent pain; if not so followed by evil consequences, the pleasure would be as pure as any other pleasure. With Bentham morality became, as it were, a question of the ledger and the balance-sheet; all feelings were reduced to the same denomination of value, and whenever we indulge in a little enjoyment, or endure a pain, the consequences in regard to subsequent enjoyment or suffering are to be inexorably scored for or against us, as the case may be. Our conduct must be judged wise or foolish according as, in the long-run, we find a favourable 'hedonic' balance-sheet.

What Mill in his earlier life thought about these foundations of the utilitarian doctrine, and the elaborate structure reared therefrom by Bentham, he has told us in his *Autobiography*, pp. 64 to 70. Subsequently Mill revolted, as we all know, against the narrowness of the Benthamist creed. While wishing to retain[1] the precision of expression, the definiteness of meaning, the contempt of declamatory phrases and vague generalities, which were so honourably characteristic both of Bentham and of his own father, James Mill, John Stuart decided to give a wider basis and a more free and 'genial' character to the utilitarian speculations.

Let us consider how Mill proceeded to give this 'genial' character to the utilitarian philosophy. It must be admitted, he says,[2] that utilitarian writers in general have placed the superiority of mental over bodily pleasures chiefly in the greater permanency, safety, uncostliness, etc., of the former —that is, in their circumstantial advantages rather than in their intrinsic nature. As regards Bentham, at least, Mill might have omitted the word *chiefly*. But according to Mill, there is no need why they should have taken such a ground.

'They might have taken the other, and, as it may be called, higher ground, with entire consistency. It is quite compatible with the principle of utility to recognise the fact, that some *kinds* of pleasure are more desirable and more valuable than others. It would be absurd, that while, in estimating all other things, quality is considered as well as quantity, the estimation of pleasures should be supposed to depend on quantity alone.'

Then Mill proceeds to point out, with all the persuasiveness of his best style, that there are higher feelings which we would not sacrifice for any quantity of a lower feeling. Few human creatures, he holds, would consent to be changed into any of the lower animals for a promise of the fullest allowance of a beast's pleasures; no intelligent human being

[1] *Autobiography*, p. 214. [2] *Utilitarianism*, p. 11.

would consent to be a fool, no instructed person would be an ignoramus, no person of feeling and conscience would be selfish and base, and so forth. Mill, in fact, treats us to a good deal of what Paley so cynically called the 'usual declamation,' on the dignity and capacity of our nature, and the worthiness of some satisfactions compared with the grossness and sensuality of others. It must be allowed that Mill has the best of it, at least with the majority of readers. Paley is simply brutal as to the way in which he depresses everything to the same level of apparent sensuality. Mill overflows with genial and noble aspirations; he hardly deigns to count the lower pleasures as worth putting in the scale; it is better, he thinks, to be a human being dissatisfied than a pig satisfied; better to be Socrates dissatisfied than a fool satisfied. If the pig or the fool is of a different opinion, it is because they only know their own side of the question. The other party to the comparison knows both sides. In the pages which follow there is much nobleness and elevation of thought. But where is the logic? We are nothing if we are not logical. But does Mill, in the fervour of his revolt against the cold, narrow restraints of the Benthamist formulas, consider the consistency and stability of his position? Let us examine in some detail the position to which he has brought himself.

It is plain, in the first place, that pleasure is with Mill the ultimate purpose of existence; for the philosophy is that of utilitarianism, and Mill distinctly assures us (*Autobiography*, p. 178) that he 'never ceased to be a utilitarian.' We must, of course, distinguish between the pleasure of the individual and the pleasure of other individuals of the race, between Egoistic and Universalistic Hedonism, as Mr. Sidgwick calls these very different doctrines. But the happiness of the race is, of course, made up of the happiness of its units, so that unless most of the individuals pursue a course ensuring happiness, the race cannot be happy in the aggregate. Now, to acquire happiness the individual must, of course, select that line of conduct which is likely to—that is, will in the majority of cases—bring happiness. He must aim at something which is capable of being reached. Mill tells us (p. 18) that if by happiness be meant a continuity of highly pleasurable excitement, it is evident enough that this is impossible to attain.

'A state of exalted pleasure lasts only moments, or in some cases, and with some intermissions, hours or days, and is the occasional brilliant flash of enjoyment, not its permanent and steady flame. Of this the philosophers who have taught that happiness is the end of life were as fully aware as those who taunt them. The happiness which they meant was not a life of rapture; but moments of such, in an existence made up of few and transitory pains, many and various pleasures, with a decided predominance of the active over the passive, and *having as the foundation of the whole, not to expect more from life than it is capable of bestowing.*[1] A life thus composed, to those who have been fortunate enough to obtain it, has always appeared worthy of the name of happiness.'

Then Mill goes on to point out what he considers has been sufficient to satisfy great numbers of mankind (p. 19)—

'The main constituents of a satisfied life appear to be two, either of which by itself is often found sufficient for the purpose: tranquillity, and excitement. With much tranquillity, many find that they can be content with very little pleasure: with much excitement, many can reconcile themselves to a considerable quantity of pain. There is assuredly no inherent impossibility in enabling even the mass of mankind to unite both.'

From these passages we must gather that at any rate the mass of mankind will attain happiness if they are satisfied with these main constituents, and we are especially told that the foundation of the whole utilitarian philosophy (Mill does not specify the substantive to which the adjective *whole* applies in the above quotation, but it must from the context be either 'utilitarian philosophy,' 'search for happiness,' or some closely equivalent idea) is *not to expect from life more than it is capable of bestowing.*

The question, then, may fairly arise whether upon a fair

[1] Italicised by the present writer.

calculation of probabilities they are not wise, upon Mill's own showing, who aim at moderate achievements in life, so that in accomplishing these they may insure a satisfied life. This seems the more reasonable, if, as Mill elsewhere tells us, the nobler feelings are very apt to be killed off by the chilly realities of life.

'Many,' he says (p. 14) 'who begin with youthful enthusiasm for everything noble, as they advance in years sink into indolence and selfishness. But I do not believe that those who undergo this very common change, voluntarily choose the lower description of pleasure in preference to the higher, I believe that before they devote themselves exclusively to the one, they have already become incapable of the other. Capacity for the nobler feelings is in most natures a very tender plant, easily killed, not only by hostile influences, but by mere want of sustenance; and in the majority of young persons it speedily dies away if the occupations to which their position in life has devoted them, and the society into which it has thrown them, are not favourable to keeping that higher capacity in exercise. Men lose their high aspirations as they lose their intellectual tastes, because they have not time or opportunity for indulging them; and they addict themselves to inferior pleasures, not because they deliberately prefer them, but because they are either the only ones to which they have access, or the only ones which they are any longer capable of enjoying. It may be questioned whether any one who has remained equally susceptible to both classes of pleasure, ever knowingly and calmly preferred the lower; though many, in all ages, have broken down in an ineffectual attempt to combine both.'

It would seem, then, that for the mass of mankind there is small prospect indeed of achieving happiness through high aspirations. They will not have time nor opportunity for indulging them. If they look for happiness solely to such aspirations they must be disappointed, and cannot have a satisfied life; if they attempt to combine the higher and lower lives they are likely to 'break down in the ineffectual attempt.' Now, I submit that, under these circumstances, it is folly, according to Mill's scheme of morality, to aim high; it is equivalent to going into a life-lottery, in which there are no doubt high prizes to be gained, but few and far between. It is simply gambling with hedonic stakes; preferring a small chance of high enjoyment to comparative certainty of moderate pleasures. Mill clearly admits this when he says (p. 14), 'It is indisputable that the being whose capacities of enjoyment are low has the greatest chance of having them fully satisfied; and a highly endowed being will always feel that any happiness which he can look for, as the world is constituted, is imperfect.'

Although, then, 'the foundation of the whole' is not to expect from life more than it is capable of bestowing, we are actually to prefer becoming highly endowed, although we cannot expect life to satisfy the corresponding aspirations. That is to say, although seeking for happiness, we are to prefer the course in which we are approximately certain of not obtaining it.

But Mill goes on to give some explanations. He says that the highly endowed being can learn to bear the imperfections of his happiness, 'if they are at all bearable' (p. 14). This is small comfort if they happen to be *not at all bearable*, an alternative which is not further pursued by Mill. And will not this intolerable fate be most likely to befall those whose aspirations have been pitched most highly? But Mill goes on—

'They (that is, the imperfections of life or happiness?) will not make him envy the being who is indeed unconscious of the imperfections, but only because he feels not at all the good which those imperfections qualify. It is better to be a human being dissatisfied, than a pig satisfied; better to be Socrates dissatisfied, than a fool satisfied. And if the fool, or the pig, is of a different opinion, it is because they only know their own side of the question. The other party to the comparison knows both sides.'

Concerning this position of affairs the most apposite remark I can make is contained in the somewhat trite and vulgar saying, 'Where ignorance is bliss, 'tis folly to be wise.' If Socrates is pretty sure to be dissatisfied, and yet, owing to his wisdom, cannot help wishing to be Socrates, he seems to have no chance of that individual happiness which depends on being satisfied, and not expecting from life more than it is capable of bestowing. The great majority of people who do not know what it is like to be Socrates, are surely to be congratulated that they can, without scruple or remorse, seek a prize of happiness which there is a fair prospect of securing. But Mill tells us that those who choose the lower life do so 'because they only know their own side of the question. The other party to the comparison knows both sides.' Then Mill introduces a paragraph, already partially quoted, in which he allows that men often do, *from infirmity of character*, make their selection for the nearer good, though they know it to be the less valuable. Many who begin with youthful enthusiasm for everything noble, sink in later years into indolence and selfishness. The capacity for the nobler feelings is easily killed, and men lose their high aspirations because they have not time and opportunity for indulging them. I submit that, *from Mill's point of view*, these are all valid reasons why they should *not* choose the higher life. We are considering here, not those who have always been devoid of the nobler feelings, but those who have in earlier life been full of enthusiasm and high aspirations. If such men, with few exceptions, decide eventually in favour of the lower life, they are parties who *do* know both sides of the comparison, and deliberately choose not to be Socrates, with the prospect of the very imperfect happiness (probably involving short rations) which is incident to the life of Socrates.

Mill, indeed, calmly assumes that the vote goes in his own and Socrates' favour. He says (p. 15)—

'From this verdict of the only competent judges, I apprehend there can be no appeal. On a question which is the best worth having of two pleasures, or which of two modes of existence is the most grateful to the feelings, apart from its moral attributes and from its consequences, the judgment of those who are qualified by knowledge of both, or, if they differ, that of the majority among them, must be admitted as final. And there need be the less hesitation to accept this judgment respecting the quality of pleasures, since there is no other tribunal to be referred to, even on the question of quantity. What means are there of determining which is the acutest of two pains, or the intensest of two pleasurable sensations, except the general suffrage of those who are familiar with both?'

Now, were we dealing with a writer of average logical accuracy there would be considerable presumption that when he adduces evidence and claims a result in his own favour in this confident way, there would be some ground for the claim. But my scrutiny of Mill's *System of Logic* has taught me caution in admitting such presumptions in respect of his writings, and here is a case in point. He claims that the suffrage of the majority is in favour of Socrates' life, although he has admitted that the vast majority of men somehow or other elect not to be Socrates. He assumes, indeed, that this is because their aspirations have been first killed off by unfavourable circumstances; his only residuum of fact is contained in this somewhat hesitating conclusion already quoted—

'It may be questioned whether any one who has remained equally susceptible to both classes of pleasures, ever knowingly and calmly preferred the lower; though many, in all ages, have broken down in an ineffectual attempt to combine both.'

Although, then, millions and millions are continually deciding against Socrates' life, for one reason or another (and many in all ages who make the ineffectual attempt at a combination break down), Mill gratuitously assumes that they are none of them competent witnesses, because they must have lost their higher feelings before they could have descended to the lower level; then the comparatively few who do choose the higher life and succeed in attaining it are adduced as giving a large majority, or even a unanimous vote in favour of their own choice. I submit that this is a fallacy probably to be best classed as a *petitio principii;* Mill entirely begs the question when he assumes that every witness against him is an incapacitated witness, because he must have lost his capacity for the nobler feelings before he could have decided in favour of the lower.

The verdict which Mill takes in favour of his high-quality pleasures is entirely that of a packed jury. It is on a par with the verdict which would be given by vegetarians in favour of a vegetable diet. No doubt, those who call themselves vegetarians would almost unanimously say that it is the best and highest diet; but then, all those who have tried such diet and found it impracticable have disappeared from the jury, together with all those whose common sense, or scientific knowledge, or weak state of health, or other circumstances, have prevented them from attempting the experiment. By the same method of decision, we might all be required to get up at five o'clock in the morning and do four hours of head-work before breakfast, because the few hard-headed and hard-bodied individuals who do this sort of thing are unanimously of opinion that it is a healthy and profitable way of beginning the day.

Of course, it will be understood that I am not denying the moral superiority of some pleasures and courses of life over others. I am only showing that Mill's attempt to reconcile his ideas on the subject with the Utilitarian theory hopelessly fails. The few pleasant pages in which he makes this attempt (*Utilitarianism*, pp. 8-28), form, in fact, a most notable piece of sophistical reasoning. Much of the interest of these undoubtedly interesting passages arises from the kaleidoscopic way in which the standing difficulties of ethical science are woven together, as if they were logically coherent in Mill's mode of presentation. The ideas involved are as old as Plato and Aristotle. The high aspirations correspond to τὸ καλὸν of Plato. The superior man who can judge both sides of the question is the βέλτιστος ἀνήρ of Aristotle. The Utilitarian doctrine is that of Epicurus. Now, Mill managed to persuade himself that he could in twenty pages reconcile the controversies of ages.

Nor is it to be supposed that Bentham, in making his analysis of the conditions of pleasure, overlooked the difference of high and low; he did not overlook it at all—he analysed it. A pleasure to be high must have the marks of intensity, length, certainty, fruitfulness, and purity, or of some of these at least; and when we take Altruism into account, the feelings must be of wide extent—that is, fruitful of pleasure and devoid of evil to great numbers of people. It is a higher pleasure to build a Free Library than to establish a new Race Course; not because there is a *Free-Library-building emotion*, which is essentially better than a *Race-Course-establishing emotion*, each being a simple unanalysable feeling; but because we may, after the model of inquiry given by Bentham, resolve into its elements the effect of one action and the other upon the happiness of the community. Thus, we should find that Mill proposed to give 'geniality' to the Utilitarian philosophy by throwing into confusion what it was the very merit of Bentham to have distinguished and arranged scientifically. We must hold to the dry old Jeremy, if we are to have any chance of progress in Ethics. Mill, at some 'crisis in his mental history,' decided in favour of a genial instead of a logical and scientific Ethics, and the result is the mixture of sentiment and sophistry contained in the attractive pages under review.

In order to treat adequately of Mill's ethical doctrines it would no doubt be necessary to go on to other parts of the Essays, and to inquire how he treats other moral elements, such as the Social or Altruistic Feelings. The existence of such feelings is admitted on p. 46, and, indeed, insisted on as a basis of powerful natural sentiment, constituting the strength of the Utilitarian morality. But it would be an endless work to examine all phases of Mill's doctrines, and to show whether or not they are logically consistent *inter se*. They are really not worth the trouble. Just let us notice, however, how he treats the question whether moral feelings are innate or not. On this point Mill gives (p. 45) the following characteristic deliverance:—'If, as is my own belief, the moral feelings are not innate, but acquired, they are not for that reason the less natural. It is natural to man to speak, to reason, to build cities, to cultivate the ground, though these are acquired faculties. The moral feelings are not indeed a part of our nature, in the sense of being in any perceptible degree present in all of us; but this, unhappily, is a fact admitted by those who believe the most strenuously in their transcendental origin. Like the other acquired capacities above referred to, the moral faculty, if not a part of our nature, is a natural outgrowth from it; capable, like them, in a certain small degree, of springing up spontaneously; and susceptible of being brought by cultivation to a high degree of development.' If life were long enough, I should like, with the assistance of the *Methods of Ethics*, to analyse the ideas involved in this passage. I can merely suggest the following questions :—If acquired capacities are equally natural with those not acquired, what is the use of introducing a distinction without a difference? If moral feelings can spring up spontaneously, even in the smallest degree, and then be developed by ' natural outgrowths,' how do any of our feelings differ from natural ones? What does Mill mean, at the top of the next page, by speaking of ' moral associations which are wholly of artificial creation?' Are these also not the less natural because they are of artificial creation? If not, we should like to know how to draw the line between *acquired* and *artificial* capacities. How, again, are we to interpret the use of the word *natural*, on p. 50, where, speaking of the deeply-rooted conception which every individual even now has of himself as a social being, he says—

'This feeling in most individuals is much inferior in strength to their selfish feelings, and is often wanting altogether. But to those who have it, it possesses all the characters of a natural feeling. It does not present itself to their minds as a superstition of education, etc.'

Here a natural feeling is contrasted to the product of education, although we were before told that acquired capacities, like speaking, building, cultivating, were none the less natural. But I must candidly confess that when Mill introduces the words *nature* and *natural*, I am completely baffled. I give it up. I can no longer find any logical marks to assist me in tracking out his course of thought. The word *nature* may be Mill's key to a profound philosophy; but I rather think it is the key to many of his fallacies.

I often amuse myself by trying to imagine what Bentham would have said of Benthamism expounded by Mill. Especially would it be interesting to hear Bentham on Mill's use of the word 'natural.' No passage in which Bentham analyses the meaning of 'nature,' or 'natural,' occurs to me, but the following is his treatment of the word 'unnatural,' as employed in Ethics:—

'Unnatural, when it means anything, means unfrequent: and there it means something; although nothing to the present purpose. But here it means no such thing: for the frequency of such acts is perhaps the great complaint. It therefore means nothing; nothing, I mean, which there is in the act itself. All it can serve to express is, the disposition of the person who is talking of it: the disposition he is in to be angry at the thoughts of it.'[1]

Would that the grand old man, as he still sits benignly pondering in his own proper bones and clothes, in the upper regions of a well-known institution, could be got to deliver himself in like style about feelings which are *not the less natural because they are acquired*.

Before passing on, however, I must point out, in the extract from p. 45, the characteristic habit which Mill has of *minimising* things which he is obliged to admit. Instead of denying straightforwardly that we have moral feelings, he says they are not present in all of us in any 'perceptible degree.' The moral faculty is capable of springing up spontaneously 'in a certain small degree.' This will remind every reader of the way in which, in his *Essays on Religion*, instead of flatly adopting Atheism or Theism, which are clear logical negatives each of the other, he concludes that though God is almost proved not to exist, He may possibly exist, and we must 'imagine' this chance to be as large as we can, though it belongs only 'to one of the lower degrees of probability.' Exactly the same manner

[1] *Principles of Morals and Legislation*, ed. 1823, vol. i, p. 31.

of meeting a weighty question will be discovered again in his demonstration of the non-existence of necessary truths. I shall hope to examine carefully his treatment of this important part of philosophy on a future occasion. We shall then find, I believe, that his argument proves non-existence of such things as necessary truths, because those truths which cannot be explained on the association principle are very few indeed. I beg pardon for introducing an incongruous illustration, but Mill's manner of minimising an all-important admission often irresistibly reminds me of the young woman who, being taxed with having borne a child, replied that it was only a very small one.

Such are the intricacies and wide extent of ethical questions, that it is not practicable to pursue the analysis of Mill's doctrine in at all a full manner. We cannot detect the fallacious reasoning with the same precision as in matters of geometric and logical science. This analysis is the less needful too, because, since Mill's Essays appeared, Moral Philosophy has undergone a revolution. I do not so much allude to the reform effected by Mr. Sidgwick's *Methods of Ethics*, though that is a great one, introducing as it does a precision of thought and nomenclature which was previously wanting. I allude, of course, to the establishment of the Spencerian Theory of Morals, which has made a new era in philosophy[1]. Mill has been singularly unfortunate from this point of view. He might be defined as the last great philosophic writer conspicuous for his ignorance of the principles of evolution. He brought to confusion the philosophy of his master, Bentham; he ignored that which was partly to replace, partly to complete it.

[1] A very important article by Dr. E. L. Youmans upon Mr. Spencer's philosophy has just appeared in the *North American Review* for October 1879. Dr. Youmans traces the history of the Evolution doctrines, and proves the originality and independence of Mr. Spencer as regards the closely related views of Mr. Darwin, Mr. Wallace, and Professor Huxley. The eminent men in question are no doubt in perfect agreement; but Dr. Youmans seems to think that readers in general do not properly understand the singular originality and boldness of Mr. Spencer's vast and partially accomplished enterprise in philosophy.

I am aware that, in her Introductory Notice to the Essays on Religion (p. viii), Miss Helen Taylor apologises for Mill having omitted any references to the works of Mr. Darwin and Sir Henry Maine 'in passages where there is coincidence of thought with those writers, or where subjects are treated which they have since discussed in a manner to which the Author of these Essays would certainly have referred had their works been published before these were written.'[1] Here it is implied that Mill anticipated the authors of the Evolution philosophy in some of their thoughts, and it is a most amiable and pardonable bias which leads Miss Taylor to find in the works of one so dear to her that which is not there. The fact is that the whole tone of Mill's moral and political writings is totally opposed to the teaching of Darwin and Spencer, Tylor and Maine. Mill's idea of human nature was that we came into the world like lumps of soft clay, to be shaped by the accidents of life, or the care of those who educate us. Austin insisted on the evidence which history and daily experience afford of 'the extraordinary pliability of human nature,' and Mill borrowed the phrase from him.[2] No phrase could better express the misapprehensions of human nature which, it is to be hoped, will cease for ever with the last generation of writers. Human nature is one of the last things which can be called 'pliable.' Granite rocks can be more easily moulded than the poor savages that hide among them. We are all of us full of deep springs of unconquerable character, which education may in some degree soften or develop, but can neither create nor destroy. The mind can be shaped about as much as the body; it may be starved into feebleness, or fed and exercised into vigour and fulness; but we start always with inherent hereditary powers of growth. The non-recognition of this fact is the great defect in the moral system of Bentham. The great Jeremy was accus-

[1] Mr. Morley does not seem to countenance any such claims. On the contrary, he remarks in his *Critical Miscellanies*, p. 324, that Mill's Essays lose in interest by not dealing with the Darwinian hypothesis.

[2] *Autobiography*, p. 187.

tomed to make short work with the things which he did not understand, and it is thus he disposes of 'the pretended system' of a moral sense—[1]

'One man says he has a thing made on purpose to tell him what is right and what is wrong, and that it is called a *moral sense;* and then he goes to his work at his ease, and says such a thing is right and such a thing is wrong—Why? because my moral sense tells me it is.'

Bentham then bluntly ignored the validity of innate feelings, but this omission, though a great defect, did not much diminish the value of his analysis of the good and bad effects of actions. Mill discarded the admirable Benthamist analysis, but failed to introduce the true Evolutionist principles; thus he falls between the two. It is to Herbert Spencer we must look for a more truthful philosophy of morals than was possible before his time.

The publication of the first part of his Principles of Morality, under the title *The Data of Ethics*, gives us, in a definite form, and in his form, what we could previously only infer from the general course of his philosophy and from his brief letter on Utilitarianism addressed to Mill. Although but fragments, these writings enable us to see that a definite step has been made in a matter debated since the dawn of intellect. The moral sense doctrine, so rudely treated by Bentham, is no longer incapable of reconciliation with the greatest happiness principle, only it now becomes a moving and developable moral sense. An absolute and unalterable moral standard was opposed to the palpable fact that customs and feelings differ widely, and Paley, on this ground, was induced to reject it. Now we perceive that we all have a moral sense; but the moral sense of one individual, and still more of one race, may differ from that of another individual or race. Each is more or less fitted to its circumstances, and the best is ascertained by *eventual success*.

At the tail end of an article it is, of course, impossible

[1] *Principles of Morals* etc. p. 29.

to discuss the grounds or results of the Spencerian philosophy. To me it presents itself, in its main features, as unquestionably true; indeed, it is already difficult to look back and imagine how philosophers could have denied of the human mind and actions what is so obviously true of the animal races generally. As a reaction from the old views about innate ideas, the philosophers of the eighteenth century wished to believe that the human mind was a kind of *tabula rasa*, or *carte blanche*, upon which education could impress any character. But if so, why not harness the lion, and teach the sheep to drive away the wolf? If the moral, not to speak of the physical characteristics of the lower animals, are so distinct, why should there not be moral and mental differences among ourselves, descending, as we obviously do, from different stocks with different physical characteristics? Notice what Mr. Darwin says on this point—

'Mr. J. S. Mill speaks, in his celebrated work, *Utilitarianism* (1864, p. 46), of the social feelings as a "powerful natural sentiment," and as "the natural basis of sentiment for utilitarian morality;" but on the previous page he says, "if, as is my own belief, the moral feelings are not innate, but acquired, they are not for that reason less natural." It is with hesitation that I venture to differ from so profound a thinker, but it can hardly be disputed that the social feelings are instinctive or innate in the lower animals; and why should they not be so in man? Mr. Bain and others believe that the moral sense is acquired by each individual during his lifetime. On the general theory of evolution this is at least extremely improbable.'[1]

Many persons may be inclined to like the philosophy of Spencer no better than that of Mill. But, if the one be true and the other false, liking and disliking have no place in the matter. There may be many things which we cannot possibly like; but if they are, they are. It is possible that the Principles of Evolution, as expounded by Mr. Herbert

---

[1] *The Descent of Man, and Selection in Relation to Sex*, 1871, vol. i, p. 71. I cannot help thinking that Mr. Darwin felt the inconsistency and confusion of ideas in the passages quoted, although he does not so express himself. Otherwise, why does he quote from two pages?

Spencer, may seem as wanting in 'geniality' as the formulas of Bentham. There is nothing genial, it must be confessed, about the mollusca and other cold-blooded organisms with which Mr. Spencer perpetually illustrates his principles. Heaven forbid that any one should try to give geniality to Mr. Spencer's views of ethics by any operation comparable to that which Mill performed upon Benthamism.

Nevertheless, I fully believe that all which is sinister and ungenial in the Philosophy of Evolution is either the expression of unquestionable facts, or else it is the outcome of misinterpretation. It is impossible to see how Mr. Spencer, any more than other people, can explain away the existence of pain and evil. Nobody has done this; perhaps nobody ever shall do it; certainly systems of Theology will not do it. A true philosopher will not expect to solve everything. But if we admit the patent fact that pain exists, let us observe also the tendency which Spencer and Darwin establish towards its *minimisation*. Evolution is a striving ever towards the better and the happier. There may be almost infinite powers against us, but at least there is a deep-laid scheme working towards goodness and happiness. So profound and widespread is this confederacy of the powers of good, that no failure and no series of failures can disconcert it. Let mankind be thrown back a hundred times, and a hundred times the better tendencies of evolution will reassert themselves. Paley pointed out how many beautiful contrivances there are in the human form, tending to our benefit. Spencer has pointed out that the Universe is one deep-laid framework for the production of such beneficent contrivances. Paley called upon us to admire such exquisite inventions as a hand or an eye. Spencer calls upon us to admire a machine which is the most comprehensive of all machines, because it is ever engaged in inventing beneficial inventions *ad infinitum*. Such at least is my way of regarding his Philosophy.

Darwin, indeed, cautions us against supposing that natural selection always leads towards the production of higher and happier types of life. Retrogression may result as well as progression. But I apprehend that retrogression can only occur where the environment of a living species is altered to its detriment. Mankind degenerates when forced, like the Esquimaux, to inhabit the Arctic regions. Still in retrograding, in a sense, the being becomes more suited to its circumstances—more capable therefore of happiness. The inventing machine of Evolution would be working badly if it worked otherwise. But, however this may be, we must accept the philosophy if it be true, and, for my part, I do so without reluctance.

According to Mill, we are little self-dependent gods, fighting with a malignant and murderous power called Nature, sure, one would think, to be worsted in the struggle. According to Spencer, as I venture to interpret his theory, we are the latest manifestation of an all-prevailing tendency towards the good—the happy. Creation is not yet concluded, and there is no one of us who may not become conscious in his heart that he is no Automaton, no mere lump of Protoplasm, but the Creature of a Creator.

V

METHOD OF DIFFERENCE

THERE are statements in Mill's logic so strangely wrong that it must be lasting matter of wonder that they had not long since opened the eyes of readers to the true character of the whole system. Among the strangest of these things is the power and importance which he attributes to his favourite Method of Difference. Many thousands of readers have studied this method. How many have discovered that in his exposition of it, Mill confuses an *experiment* with the *generalisation* founded on the experiment; that is to say, he actually confuses a material operation, or at best the observation of what takes place in a material operation, with the reasoning which leads us to infer that what takes place in one experiment will also take place in another experiment. He comes to the astonishing conclusion that a general law of nature may be founded upon the observation of two instances. The same philosopher who formerly insisted that induction was strictly speaking from particulars to particulars, discovers that provided the requisitions of the Method of Difference be fulfilled, two particular instances give us at once the universal law, and this assertion applies to any kind of matter of fact.

But let us consider carefully how Mill describes and illustrates this Method of Difference. He says [1]—

[1] Chap. viii, sec. 2, beginning.

'In the Method of Agreement we endeavoured to obtain instances which agreed in the given circumstances but differed in every other: in the present method we require, on the contrary, two instances resembling one another in every other respect, but differing in the presence or absence of the phenomenon we wish to study. If our object be to discover the effects of an agent A, we must procure A in some set of ascertained circumstances, as A B C, and having noted the effects produced, compare them with the effect of the remaining circumstances B C, when A is absent. If the effect of A B C is $a\ b\ c$, and the effect of B C, $b\ c$, it is evident that the effect of A is a.'

After pointing out how we may begin at the other end and selecting an instance in which a occurs, such as $a\ b\ c$, must look out for another instance in which the remaining circumstances $b\ c$ occur without a, Mill proceeds to illustrate the use of the method in these words [1]—

'It is scarcely necessary to give examples of a logical process to which we owe almost all the inductive conclusions we draw in early life. When a man is shot through the heart, it is by this method we know that it was the gun-shot which killed him: for he was in the fulness of life immediately before, all circumstances being the same, except the wound.'

Briefly described then, this method requires that all circumstances remain exactly the same except one particular circumstance, which being altered may be regarded as the cause of whatever alteration appears in the consequents. A man is in the fulness of health; a shot passes through his heart; he falls dead; *ergo* the gun-shot is the cause of his falling dead; this is approximately certain, because it is very unlikely that any other circumstance should have been altered, that any other cause but the shot should have come into operation just at the same moment. But then there is no generalisation in this experiment; it is the mere observation of a man in one state followed by a shot and the man in an altered state. Let us see what Mill thinks may be learnt from one experiment.

In a number of places he asserts in the most unequivocal

[1] Next paragraph.

way that two instances are sufficient to give a general law provided they conform to the requirements of the Method of Difference. I do not, indeed, find this explicitly stated where it ought to be stated, namely, in the formal description of the Method; but later on we find such passages as the following:[1]—

'Plurality of causes, therefore, not only does not diminish the reliance due to the Method of Difference, but does not render a greater number of observations or experiments necessary: two instances, the one positive and the other negative, are still sufficient for the most complete and rigorous induction.'

In the twenty-first chapter of the third book,[2] speaking of the Method of Difference, he says—

'That Method authorises us to infer a general law from two instances; one in which A exists together with a multitude of other circumstances, and B follows; another, in which, A being removed, and all other circumstances remaining the same, B is prevented. What, however, does this prove? It proves that B, in the particular instance, cannot have had any other cause than A; but to conclude from this that A was the cause, or that A will on other occasions be followed by B, is only allowable on the assumption that B must have some cause; that among its antecedents in any single instance in which it occurs, there must be one which has the capacity of producing it at other times. This being admitted, it is seen that in the case in question that antecedent can be no other than A; but, that if it be no other than A it must be A, is not proved, by these instances at least, but taken for granted.

These remarks can bear no other construction than that what takes place in the instances examined will take place always; that is, where A is present, *a* will follow. The induction, Mill says, is rigorous and complete, and induction, we remember, was inference from known to unknown cases. It follows, for instance, that because one man was shot through the heart and died, therefore all men shot through

[1] Book iii, chap. ix, sec. 2, middle of this paragraph.
[2] Middle of first paragraph.

the heart will die. Surely this is an obvious inference and it gives a general law—All men shot through the heart die. Does the reader think that the Method of Difference authorises such a conclusion? If so, the results are alarming. Because one man pricks his finger and dies as a result of it, all men pricking their fingers must die? Because one man eats an egg and is immediately ill, all men must be made ill by an egg? Because one man goes to sea and suffers from nausea, all men must suffer likewise? Such ridiculous results are the outcome of Mill's Method. A man is in the fulness of health; he steps on a boat which pushes off into a rough sea; he suffers from nausea; no circumstances are changed except the dry land for the heaving boat; therefore A the heaving boat must be the cause of a the nausea. So far good; but Mill says we have a rigorous and complete induction, and may draw a general conclusion; this can be no other than that all men stepping on to a heaving boat will suffer the effect a.

There is no difficulty in seeing where Mill's blunder lies. He has overlooked the multiplicity of circumstances, and that a cause is usually if not always a conjunction of many circumstances. Nothing can be more opposed to fact and science than that one definite cause A is joined with one definite effect a. A man steps on to a heaving boat; no doubt this is a definite antecedent which may be symbolised by A, if you like, and there is a definite result a; but then the other antecedents are almost innumerable—the man's constitution, his habit of travelling at sea or otherwise, his state of health at the time, the accidental condition of his stomach, his posture, or even the way in which his thoughts are engaged.

The fact obviously is, that from a single experiment all we can infer is, that the like result will follow in like circumstances. If the same man with the same state of health and other circumstances steps on to a heaving boat, he will be ill. If Thomas Jones dies from the prick of a pin, then William Roberts will also die from a like cause, provided he be in exactly the same state of health and in all material circumstances the same as Thomas Jones. But such inferences are not general laws; they are at the best to be called particular inferences; in fact, as a general rule, the inference would not be warranted at all.

Mill's example of the man shot through the heart may now be seen to be thoroughly misleading. He gives it as an instance of inductive conclusion, and no doubt it is true that a man shot through the heart does in all probability die in consequence of the shot. Moreover, all men shot through the heart will doubtless die; but does any one venture to say that one single experiment of the kind proves that all men shot through the heart will die? If so, because one man has his leg cut off and dies, all men who have their legs cut off will die. It is obvious that in the case of the gun-shot, we have a great deal of knowledge about the heart and its functions in sustaining life. We believe that whenever the heart is disabled death must result, but not on the ground of any single experiment.

THE END